CW01573129

CATALOG

OF

RELIGIOUS ARTICLES

AND

Lithographs, Engravings, and Lace Pictures.

CONTENTS.

RELIGIOUS ARTICLES.

RELIGIOUS PICTURES.

MISSIONS SUPPLIED AT LOWEST RATES.

BENZIGER BROTHERS, NEW-YORK AND CINCINNATI.

GOLD AND SILVER MEDALS.

Our Gold Medals are 18 Carats fine

EDALS WITH THE IMMACULATE CONCEPTION ON ONE SIDE, AND THE NAME OF MARY AND SACRED HEARTS OF JESUS AND MARY ON REVERSE.

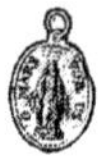

No. 0.

Per doz.
Gold, $
Silver,

No. 1.

Per doz.
In Gold, $5.00
In Silver, 31

No. 2.

Per doz.
In Gold, $6.60
In Silver, 43

No. 3.

Per doz.
In Gold, $11.50
In Silver, 81

No. 4.

Per doz.
In Gold, $17.00
In Silver, 1.25

No. 5.

Per doz.
In Gold, $26.00
In Silver, 1.85

No. 6.

Per doz.
Gold, $37.00
Silver, 2.45

No. 7.

Per doz.
In Gold, $46.00
In Silver, 3.35

No. 8.

Per doz.
In Gold, $
In Silver, 4.60

No. 9.

Per doz.
In Gold, $
In Silver, 9.70

No. 10.—1¼ inch diam. long, { In Gold, per doz., $
In Silver, " 11.25

No. 15.

Reverse: St. Joseph.
Per doz.
In Gold, $26.00
In Silver, 1.85

No. 17.

Reverse: Imm. Concept.
Per doz.
In Gold, $26.00
In Silver, 1.85

No. 18.

Reverse: Imm. Concept.
Per doz.
In Gold, $26.00
In Silver, 1.85

No. 26.

Reverse: Name of Mary.
Per doz.
In Gold, $17.00
In Silver, 1.25

PRICES IN CURRENCY.

BENZIGER BROTHERS, NEW-YORK AND CINCINNATI.

No. 19.

Reverse:
Guardian Angel.

Per doz.
In Gold, $46.00
In Silver, 3.35

No. 35.

Reverse:
St. Alphonsus Liguori.

Per doz.
In Gold, $40.00
In Silver, 4.50

No. 39.

Reverse:
Sacred Heart of Jesus.

Per doz.
In Gold, $54.00
In Silver, 4.50

No. 25.

Reverse:
Immaculate Conception

Per doz.
In Gold, $60.00
In Silver, 4.60

No. 40.

Reverse:
Our Lady of Lourdes.

Per doz.
In Gold, $46.00
In Silver, 3.35

No. 45.
The same Medal.
Size of No. 16.

Per doz.
In Gold, $26.00
In Silver, 1.85

No. 16.

Reverse:
Guardian Angel.

Per doz.
In Gold, $26.00
In Silver, 1.85

No. 20.
The same Medal.
Size of No. 37.

Per doz.
In Gold, $46.00
In Silver, 3.35

No. 37.

Reverse:
Sacred Heart of Mary

Per doz.
In Gold, $46.00
In Silver, 3.35

No. 36.
The same Medal.
Size of No. 16.

Per doz.
In Gold, $26.00
In Silver, 1.85

No. 38.

Reverse:
Immaculate Conception.

Per doz.
In Silver, $16.00

No. 27.—1 inch diameter long; Holy Communion; Reverse: Remembrance of first Communion, with space for inscription; in Silver, per doz., . . . $10.0

No. 41.—⅞ inch diameter long; Congregation of the Children of Mary; Reverse: Immaculate Conception; in Silver, per doz., ... 3.3

Most of the above Medals can be had with scalloped edges.

Prices of Gold Medals sometimes vary a little according to their weight.

PRICES IN CURRENCY.

BRASS MEDALS.

os.	Long Diam.			Per gross.
0	5/16 inch,	Immaculate Conception; reverse: Name of Mary, and Sacred Hearts of Jesus and Mary.	White.	$0 30
1	3/8 "	" " "	White.	32
2	7/16 "	" " "	White.	34
3	9/16 "	" " "	White.	36
4	5/8 "	" " "	White.	44
4	5/8 "	" " "	Yellow.	60
5	11/16 "	" " "	White.	56
5	11/16 "	" " "	Yellow.	72
6	13/16 "	" " "	White.	77
6	13/16 "	" " "	Yellow.	93
7	7/8 "	" " "	White.	96
7	7/8 "	" " "	Yellow.	1 20
8	1 "	" " "	White.	1 20
9	1 1/8 "	" " "	White.	2 05
0	1 1/4 "	" " "	White.	4 10
1	1 3/8 "	" " "	White.	5 30
2	1 1/2 "	" " "	White.	8 00
3	1 11/16 "	" " "	White.	21 00
4	1 7/8 "	" " "	White.	29 00
5	11/16 "	St. Joseph; reverse: Guardian Angel.	"	56
6	11/16 "	St. Aloysius; " " "	"	56
8	11/16 "	Congregation of the Holy Child Jesus; reverse: Immac. Concep.	"	56
9	7/8 "	St. Joseph; reverse: Guardian Angel.	"	96
20	7/8 "	St. Aloysius; " " "	"	96
21	7/8 "	Mater Dolorosa; reverse: 1080 days' indulgence.	"	96
22	7/8 "	St. Ann and Mary; " Sacred Hearts of Jesus and Mary	"	96
23	7/8 "	Ecce Homo; " Mater Dolorosa.	"	96
24	7/8 "	Holy Family; " Guardian Angel.	"	96
25	7/8 "	(round) Congregation of the Children of Mary.	"	1 20
28	7/8 "	St. Benedict; reverse: Cross of St. Benedict. Yellow, $1.20;	"	96
30	1 3/8 "	Mater Dolorosa; " 1080 days' indulgence.	"	5 30
31	1 3/8 "	Seven Dolors; " Crucifixion.	"	5 30
32	1 3/8 "	Holy Family; " Sacred Hearts of J., M., and J., Yellow, $8.70;	"	5 30
34	3/4 "	(round) Mater Admirabilis; reverse: Sacred Heart of Mary.	White.	96
37	7/8 "	Sacred Heart of Jesus; " " " "	"	96
42	11/16 "	St. Ann and Mary; reverse: Sacred Hearts of Jesus and Mary.	"	56
43	1 1/8 "	Congregation of the Children of Mary.	"	2 05
44	1 1/4 "	(round) Mission Medal: Miraculous Virgin; reverse: Name of Jesus,	"	4 10
01	1 1/2 "	Miraculous Virgin; reverse: Name of Jesus, and Sacred Hearts.	Yellow.	8 10
02	1 1/2 "	Ecce Homo; " Mater Dolorosa.	"	8 10
06	1 1/4 "	St. Joseph; " Guardian Angel.	"	5 40
07	1 1/4 "	St. Ann; " Sacred Hearts.	"	5 40
08	1 1/4 "	St. Aloysius; " Immaculate Conception.	"	5 40
10	1 1/4 "	Pope Pius IX.; " S. S. Peter and Paul.	"	5 40
11	1 1/4 "	Immaculate Concept.; " Name of Jesus, and Sacred Hearts.	"	5 40
12	1 1/4 "	St. Benedict; " Cross of St. Benedict.	"	5 40
13	1 1/4 "	St. Francis of Assis.; " Miraculous Virgin.	"	5 40
7	1 1/8 "	Sacred Heart of Jesus; " Sacred Heart of Mary.	"	5 40
8	1 1/8 "	St. Joseph; " Immaculate Conception.	"	4 05
19	1 1/8 "	St. Ignatius; " Name of Jesus, and Sacred Hearts.	"	2 70
20	1 1/8 "	St. Stanislaus; " St. Aloysius.	"	2 70
21	1 1/8 "	Ecce Homo; " Mater Dolorosa.	"	2 70
22	1 1/8 "	Miraculous Virgin; " Name of Jesus, and Sacred Hearts.	"	2 70
27	13/16 "	(form of a Heart,) Sacred Hearts of Jesus and Mary. White or	"	1 85

PRICES IN CURRENCY.

SILVER MEDALLIONS AND RELIQUARIES.

SILVER MEDALS.

BETWEEN GLASS.

Nos. 201 & 202.
Per doz. 74c.
GILT.
Per doz. $1.40.

Nos. 203 & 204.
Per doz. $1.05.
GILT.
Per doz. $2.45.

Nos. 205 & 206.
Per doz. $2.15.
GILT.
Per doz. $4.60.

MEDALLIONS.

ALL SILVER.

No. 207.
Per doz. $1.80.

No. 208.
Per doz. $2.45.

SILVER MEDALLIONS.

WITH GLASS ON ONE SIDE.

No. 209.
Per doz. $2.05.

No. 210.
Per doz. $2.30.

No. 211.
Per doz. $1.90.

SILVER MEDALLIONS.

WITH GLASS ON BOTH SIDES.

No. 212.
Per doz. $2.85.

No. 213.
Per doz. $3.55.

RELIQUARIES.

ALL SILVER.

No. 214.
Per doz. $3.80.

No. 215.
Per doz $4.35.

No. 216.
Per doz. $4.90.

No. 217.
Per doz. $6.50.

No. 218.
Per doz. $4.35.

No. 219.
Per doz. $4.05.

No. 220.
Per doz. $3.55.

PRICES IN CURRENCY.

BENZIGER BROTHERS, NEW-YORK, AND CINCINNATI, O.

SILVER RELIQUARIES

WITH GLASS COVER.

No. 221.
Per doz. $2.05.

No. 222.
Per doz. $3.55.

No. 223
Per doz. $3.00.

No. 224.
Per doz. $4.35.

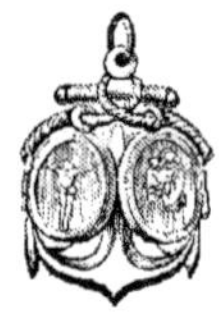

No. 225.
Per doz. $5.40.

SILVER RELIQUARY

WITH GLASS ON

ONE SIDE. BOTH SIDES.

No. 226.
'er doz. $5.70.

No. 227.
Per doz. $1.05.

SILVER MEDALLIONS

WITH RELIGIOUS PHOTOGRAPHIC VIEWS.

No. 228.
Per doz. $3.80.

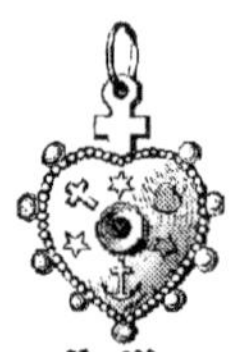

No. 229.
Per doz. $5.20.

No. 230.
Per doz. $3.95.

FAITH, HOPE & CHARITY.

No. 231.
'er doz. $2.20.

SILVER RING

OF THE
15 MYSTERIES

No. 232.
Per doz. $2.50.

SILVER MEDALS

FOR REWARD OF MERIT.

No. 233.
Small, per doz. $6.90.
Large, " " 8.30.

No. 234.
Small, per doz. $12.25.
Large, " " 23.00.

GOLD LOCKETS

ENAMELED, WITH RELIGIOUS DESIGNS.

From $3.50 to $5.

PRICES IN CURRENCY.

GOLD CROSSES.

(18 Carats.)

No. 1.
Per doz., $10.

No. 2.
Per doz., $18.

No. 3.
Per doz., $24.

No. 4.
Per doz., $11.

No. 5.
Per doz., $15.

No. 6.

WITH PEARL IN CENTRE.

Per doz., $21.

No. 7.

WITH RED OR BLUE IMITATION CORAL STEM AND GOLD LEAVES.

Per doz., $18.

SILVER CROSSES.

SOLID.
No. 23.
Per doz., $5.

No 21.
Per doz., $1.15.

SOLID.
No. 22.
Per doz , $3.

No. 24.
Per doz., $4.10.

SILVER RELIQUARY CROSSES.

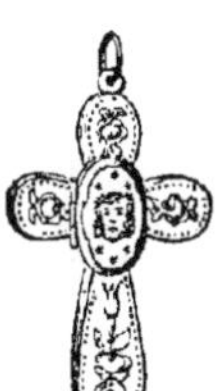

No. 25.
Per doz., $3.60.

No. 26.
Per doz., $3.85.

No. 27.
Per doz., $4.60.

No. 28.
Per doz., $4.35

We also keep a great variety of larger Gold and Silver Crosses.

PRICES IN CURRENCY.

BENZIGER BROTHERS, NEW-YORK AND CINCINNATI.

PEARL CROSSES WITH SILVER FIGURES.

ROUND.
No. 30.

	Length.	Per doz.
No. 30,	1½ inch.,	$4.25
" 31,	2 "	7.70

WITH SILVER TIPS.
No. 33.

	Length.	Per doz.
No. 32,	2 inch.,	$11.25
" 33,	2¼ "	15.50
" 34,	2¾ "	24.00

WITH SILVER TIPS.
No. 35.

	Length.	Per doz.
No. 35,	1¾ inch.,	$10.25

Nos. 37 and 39.

WITH SILVER TIPS.

No. 36, 1½ inch. long, per doz.,	$6 90
" 37, 2 " " "	11 75
" 38, 2½ " " "	17 50

WITH SILVER BACK AND TIPS.

No. 39, 1¾ inch. long, per doz.,	$16 50
" 40, 2¼ " " "	22 00
" 41, 2¾ " " "	31 00
" 42, 3¼ " " "	42 00

SILVER-BOUND.

No. 43, 1⅛ inch. long, per doz.,	$6 90
" 44, 1½ " " "	8 50
" 45, 2¼ " " "	17 50

os. 43, 46, and 50a.

COCOA CROSSES.

SILVER-BOUND.

No. 46, 1⅛ inch. long, per doz.,	$5 30
" 47, 1½ " " "	6 40
" 48, 2 " " "	9 60
" 49, 2¼ " " "	12 25

EBONY CROSSES.

SILVER-BOUND.

No. 50a, 1⅛ inch. long, per doz.,	$5 30
" 50b, 1½ " " "	6 40
" 50c, 2 " " "	9 60
" 50d, 2¼ " " "	12 25

PRICES IN CURRENCY.

BRASS-BOUND CROSSES.

No. 56.

WITH BRASS BACKS.

			Per doz.
No. 51,	7/8 inch. long,		$0 34
" 52,	1	" "	42
" 53,	1 1/4	" "	53
" 54,	1 1/2	" "	64
" 55,	1 3/4	" "	75
" 56,	2	" "	83
" 57,	2 1/2	" "	94
" 58,	3	" "	1 35
" 59,	3 1/8	" "	1 70
" 60,	3 1/2	" "	2 05
" 61,	4	" "	2 50
" 62,	4 1/2	" "	3 15
" 63,	5	" "	4 50

No. 71.

WITH WOODEN BACKS.

			Per doz.
No. 71,	2 1/2 inch. long,		$1 35
" 72,	3	" "	1 80
" 73,	4 1/8	" "	3 15
" 74,	4 1/2	" "	3 60
" 75.	5 1/8	" "	4 95
" 76,	5 7/8	" "	5 90
" 77,	6 1/8	" "	6 30
" 78,	10 1/4	" "	23 00

No. 96.

RELIQUARY CROSSES,

WITH WOODEN BACKS,

BURNISHED.

			Per doz
No. 96,	2 1/8 inch. long,		$3 40
" 97,	3 1/4	" "	5 90

WITH WOODEN BACKS, BURNISHED.

No. 81,	6 1/2 inch. long,	per doz.,			$9 00
" 82,	8	"	"	"	13 50
" 83,	11 1/4	"	"	"	41 00

WITH WOODEN BACKS AND HEAVY BRASS TIPS, BURNISHED.

No. 91,	4 1/2 inch. long,	per doz.,			$5 90
" 92,	5 1/8	"	"	"	7 70
" 93,	6 5/8	"	"	"	10 00

CHEAP BRASS-BOUND CROSSES.

No. 121,	1	inch. long, per doz., with brass backs,	$0 32
122,	1 1/4	" " " " "	34
123,	1 1/2	" " " " "	50
124,	1 5/8	" " " " "	63
125,	1 7/8	" " " " "	68
126,	2 1/2	" " " " "	79
127,	2 3/4	" " " " "	95
128,	3	" " " " "	1 15
129,	3 1/2	" " " " "	1 60

PRICES IN CURRENCY.

BRASS CROSSES.

Nos.	Length.	Per Gross.	Nos.	Length.	Per Gross.
501	1 inch. white or yellow,	$0 51	510	1 7/8 inch. white,	$2 10
502	1 " "	92	511	2 " " or yellow,	2 45
503	1 1/8 " "	1 05	512	2 1/4 " "	3 00
504	1 3/8 " "	1 35	513	2 5/8 " yellow,	4 25
505	1 1/2 " " "	1 45	514	3 1/4 " "	9 50
506	1 1/2 " "	1 45	515	3 3/4 " "	13 50
507	1 1/2 " "	1 45	516	4 1/4 " "	13 50
508	1 1/2 " "	1 70	517	5 3/8 " "	35 00
509	1 5/8 " " "	1 85			

ROSARY CASES.

Nos.		Per doz.
1.	Wood, egg shape,	from $0 40 to $0 80
3.	" apples, pears, etc.,	" 85 " 1 20
5.	Cocoa, white, polished,	" 1 50 " 2 50
7.	" " carved,	" 2 50 " 3 00
9.	" assorted colors, carved,	" 3 00 " 6 00
11.	Cocoa, brown, carved, egg-shape,	1 40
12.	" fine " "	1 75
13.	" " " extra large egg-shape,	2 05
14.	" " " apples and pears,	1 50
15.	" extra fine " different shapes,	2 05
16.	Pearl shell, white, silver-plated mountings,	2 75
17.	" " " "	3 85
18.	" " " "	5 50
19.	" " " "	6 60
20.	" " " "	7 70
21.	" crimson, " "	3 85
22.	" " " "	4 95
23.	" " " "	6 60
24.	" " " "	7 70
25.	" " " "	8 80
26.	" spotted, " "	5 50
27.	" " " "	6 60
28.	" " " "	7 20
29.	" white, " " lined with silk,	4 15
30.	" " " " " "	4 70
31.	" " " " " " silver-plated statue inside,	13 25
32.	" crimson, " " " "	6 60
33.	" " " " " "	8 80
34.	" spotted, winding, silver-plated mountings,	6 60
35.	Porcelain shell, spotted,	6 60

PRICES IN CURRENCY.

ROSARIES.

WOODEN BEADS.

Nos.		Length.		Per doz
301,	Black, brown, or maroon, steel chain,	11	inches	$0 3
302,	" " " " "	13	"	3
303,	" " " " "	15½	"	3
304,	" " " " " and cross,	12½	"	4
305,	" " " " " "	15	"	5
306,	" " " " " "	16½	"	6
307,	" " " " " "	15	"	65
308,	" " " " " "	16	"	7
309,	" " " " " decades and cross,	14½	"	8
310,	" " " extra long steel chain and cross,	16½	"	7
311,	" " " " " "	19	"	8
312,	" " " " " and decades and cross,	18	"	9
313,	" brass chain,	16	"	1 0
314,	" " "	19	"	1 2
315,	" " "	23	"	1 7
316,	" " " and brass-bound crosses 2 inches long,	19	"	1 9
317,	" " " " " " 2½ " "	23	"	2 3
318,	" or brown, steel chain,	14½	"	91
319,	" " " "	16½	"	1 0
320,	" " " "	21	"	1 5
321,	" " silver-plated chain and cross,	14	"	9
322,	" " " " "	16	"	1 1
323	" " " " " and decades,	12	"	1 1
324,	" " " " " "	14	"	1 4
331,	Black polished, steel chain,	12	"	4
332,	" " " "	14	"	5
333,	" " " " and cross,	15	"	5
334,	" " silver-plated chain and cross,	13	"	7
335,	" " " " " "	14½	"	97
336,	" " brass chain,	17	"	1 2
337,	" " "	18	"	1 5
338,	" " " and brass-bound crosses, 2 inches long,	17	"	1 9
339,	" " " " " " 2½ " "	18	"	2 7

BEADS OF BONE.

Nos.		Length.		Per doz
345,	White, silver-plated chain and cross,	11	inches,	$0 9
346,	" " " " steel cross,	13½	"	1 2
347,	" " " " cross,	12½	"	1 3
348,	" " " " "	13½	"	1 5
349,	Assorted colors, silver-plated chain and cross,	14	"	1 6
350,	Red, steel-chain and cross,	11	"	1 0
351,	" " "	14	"	1 3
352,	" carved, silver-plated cross,	14	"	2 1

PRICES IN CURRENCY.

BEADS OF GLASS.

Nos.			Length.	Per doz.
354,	White or blue, brass chain, ...	For Children.	6½ inches,	$
355,	Black, steel chain, ...		9½ "	31
356,	Red, white, blue, or green; silver-plated chain and cross, .		9 "	41
357,	" " " " " " " " .		11½ "	55
358,	" " " " " " " " . .		12 "	74
361,	Imitation of Cornelian, red; silver-plated chain and cross, .		13 "	95
362,	" " white or blue; silver-plated chain and cross,		12½ "	1 25
363,	" " " " " " " "		14½ "	1 40
365,	Brilliant, white or blue; silver-plated chain and cross, . . .		13 "	2 15
366,	" " " " " " and decades,		15 "	2 80
376,	Imitation of copper; " " "		10½ "	75
377,	" " " " " . .		12 "	90

IMITATION MOTHER-OF-PEARL BEADS.

Nos.		Length.	Per doz.
370,	White; silver-plated chain and cross, ...	12½ inches,	$1 40
371,	" or buff; silver plated chain and cross, ...	13 "	1 80
372,	" " " " " "	15 "	2 35
373,	" " " " " " and decades, .	17 "	3 05

BEADS OF STEEL.

Nos.		Length.	Per doz.
378,	Steel chain and steel cross, ...	11 inches,	$1 15
379,	" " "	13 "	1 40

BEADS OF COCOA.

Nos.		Length.	Per doz.
381,	Black, cut; steel chain and cross, ...	14 inches,	$0 79
382,	" " " " "	15 "	1 05
383,	Brown " extra long steel chain and cross, ...	16½ "	1 15
384,	Black, " " " "	19½ "	1 50
385,	" " silver-plated chain and cross, ...	12 "	1 05
386,	" " " " "	13 "	1 20
387,	" " " " "	16 "	1 50
388,	" " " " "	16½ "	1 80
389,	" " " " "	18½ "	2 25
395,	White, round, " " "	11 "	1 60
396,	" " " " "	12 "	1 85
397,	" " " " "	14 "	2 15
398,	" " " " "	14½ "	2 60
401,	" carved, " " "	12½ "	2 25
402,	" " " " "	14½ "	2 55
403,	" " " " "	15 "	3 00
406,	Brown, round, " " "	13½ "	1 55
407,	" " " " "	17 "	3 30
408,	" " extra long steel chain and cross, ...	18 "	1 55
409,	" " " " cocoa cross and figure, .	18 "	2 80
410,	" " " " and cross, ...	19 "	1 65
411,	" " " " cocoa cross and figure, .	19 "	2 90
412,	" fine carved; steel chain and cross, ...	15 "	2 20

PRICES IN CURRENCY.

Nos.		Length.	Per doz.
413,	Brown, fine carved; steel chain and cross,	16½ inches,	$2 45
414.	" " extra long steel chain and cross,	20 "	2 10
415.	" " " " cocoa cross and figure,	20 "	4 45
416.	" " steel chain and cross,	12 "	2 00
417.	" " " cocoa cross and figure,	12 "	3 20
418,	" " silver-plated chain and cross,	13 "	1 85
419,	" " " " "	14½ "	2 15
420,	" " " " cocoa cross and figure,	14½ "	3 25
421,	" " " " and cross,	18 "	3 05
422,	" " " " "	17½ "	3 20
423,	" " " " cocoa cross and figure,	18 "	4 30

IMITATION COCOA BEADS.

Nos.		Length.	Per doz.
430,	Black, brown, or maroon; silver-plated chain and cross,	10½ inches,	$0 90
431,	" " " " " "	13 "	1 05
432,	" " " " " "	16 "	1 35
433,	" " " " " "	17½ "	1 65
435,	Black or brown, carved; steel chain and cross,	13 "	95
436,	" " " " "	17 "	1 15
437,	" " " silver-plated chain and cross,	13 "	1 30
438,	" " " " " "	15 "	1 40

BEADS OF FRUITS.

Nos.		Length.	Per doz.
441,	Black; silver-plated chain and cross,	15½ inches,	$1 50
442,	Job's tears; " " "	17½ "	1 00
443,	Palmier; " " "	15 "	2 95

PERFUMED BEADS.

Nos.		Per doz.
445,	Brown or black, on elastics, with cross,	$3 10
446,	" " " "	3 95
447,	" " " "	5 10
448,	" " " "	6 80

ROSARIES OF THE SEVEN DOLORS.

Nos.		Length	Per doz.
451,	Imitation cocoa, black or brown; steel chain,	15½ inches,	$0 96
452,	Bone, white; silver-plated chain,	13 "	1 50
453,	Cocoa, black, cut; " "	14 "	1 25
454,	" " " " "	15 "	1 45

ROSARIES OF THE FIVE WOUNDS.

Nos.		Length.	Per doz.
456,	Cocoa, black, cut; silver-plated chain,	11 inches,	$0 85

ROSARIES OF THE IMMACULATE CONCEPTION.

Nos.			Length.	Per doz.
458,	Bone, white; silver-plated chain,	With an equal number of prayers, English or German.	5 inches,	$0 77
459,	Cocoa, black, cut; silver-plated chain,		6 "	74
460,	" brown, carved; " "		5 "	1 05
461,	" white, " " "		5½ "	1 00
462,	Imitation mother-of-pearl; silver-plated chain,		6 "	87

ROSARIES OF THE SACRED HEART.

Nos.		Length.	Per doz.
465,	Cocoa, brown, round,		
466,	Bone, red,		

ROSARIES FOR RELIGIOUS COMMUNITIES

Are generally kept on hand or will be made to order.

PRICES IN CURRENCY.

BENZIGER BROTHERS, NEW-YORK, AND CINCINNATI.

ROSARIES WITH SILVER CHAINS AND CROSSES.

		Per piece.
mitation Garnet.		
Silver chain and cross, ornamented decades,		from $1 50 to 3 60
'ocoa.		
Black, cut, silver chain and cross,	"	1 75 " 2 80
Brown or black, round, silver chain and cross,	"	56 " 1 85
" " extra fine carved, silver chain and cross,	"	1 00 " 2 25
White, round, " "	"	85 " 2 25
" fine carved, " "	"	1 25 " 2 75
mitation Ivory.		
Silver chain and cross,	"	75 " 1 90
Fine carved, silver chain and cross,	"	81 " 1 55
'almier (fruit).		
Silver chain and cross,	"	81 " 1 80
'eal Coral.		
No. 1, silver chain and cross,		1 25
Fine carved decades, silver chain and cross,	"	3 00 " 4 25
'eal Mother-of-Pearl.		
Fine polished, silver chain and cross.	"	1 25 " 1 85
" fine carved decades, silver chain and cross,	"	2 65 " 5 00
mitation Mother-of-Pearl.		
Fine polished, white or buff, silver chain and cross,	"	72 " 1 55
" " silver decades, silver chain and cross,	"	1 60 " 1 75
'eal Jet Cornelian.		
Fine carved decades, silver chain and cross,		" 4 55
'eal Jet, White.		
Fine carved decades, silver chain and cross,		" 6 70
'eal Jet, Onyx.		
Fine carved decades, silver chain and cross,		" 4 80
apis Lazuli		
Fine carved decades, silver chain and cross,		" 6 30
'rystal.		
Fine cut, fine carved decades, silver chain and cross,		" 11 50
mber.		
Yellow or black, silver chain and cross,		" 3 10
mitation Jet.		
White or blue, silver chain and cross,	"	66 " 1 25
" fine carved decades, silver chain and cross,		" 1 35
'erusalem Fruit.		
Red, fine carved decades, silver chain and cross,		" 1 85
'ilver Beads.		
Filigree, silver decades, silver chain and cross,	"	3 85 " 12 00

ROSARIES WITH GOLD CHAINS AND GOLD CROSSES.

'eal Lapis Lazuli, Onyx or Crystal.
Extra fine cut, with real gold chain and gold cross, per piece, . from $30 00 to $60 00

PRICES IN CURRENCY.

ROSARY BEADS ON STRINGS.

Nos.		Per doz.
1002,	Wood, black or brown.	$ 22
1003,	" "	23
1004,	" "	28
1005,	" "	35
1006,	Imitation cocoa, black or brown,	35
1007,	" "	35
1008,	" "	38
1009,	" "	38
1010,	" "	55
1011,	" "	66
1016,	" " carved,	62
1017,	" " "	62
1018,	" " "	69
1019,	" " "	84
1040,	Fruit, black,	75
1079,	Real cocoa, black, cut,	32
1080,	" " "	36
1081,	" " "	43
1082,	" " "	49
1083,	" " "	61
1084,	" " "	75
1085,	" " "	90
1092,	" black or brown, round,	75
1094,	" " "	84
1095,	" " "	90
1096,	" " "	1 10
1097,	" " "	1 65
1098,	" " "	2 20
1164,	Bone, white, round,	43
1165,	" " "	61
1166,	" " "	72
1166½,	" "	87

SCAPULARS AND SCAPULAR PRINTS.

PRINTED ON MUSLIN.

Mount Carmel (new plate), 16 scapulars on a sheet,	per 100 sheet,	$10 00
" " made up,	per dozen,	1 00
The Passion of our Lord, 16 scapulars on a sheet,	per 100 sheet,	10 00
" " made up,	per dozen,	1 00
Sacred Heart of Jesus, red print, 15 scapulars on a sheet,	per 100 sheet,	9 00
" " " made up, with prayer,	per dozen,	75
The Immaculate Conception, 16 scapulars on a sheet,	per 100 sheet,	12 00
" " made up,	per dozen,	1 00
Redemptorists' Scapulars (5 colors), made up,	"	2 30
Mount Carmel (new plate), 16 scapulars on a sheet, printed on fine silk,	per sheet,	65

Cord of St. Francis,	per dozen,	1 75
Cord of St. Thomas,		85

PRICES IN CURRENCY.

WOODEN CRUCIFIXES WITH STAMPED BRASS FIGURES.

Nos. 111-118.

Nos.		Per doz.
101, 7	inches high, round stand,	$0 49
102, 8½	" " " "	67
103, 9¾	" " " "	89
104, 11¼	" " " "	1 35
105, 13	" " " "	2 40
106, 15	" " " "	3 20
107, 19	" " " "	5 30
108, 21	" " " "	6 40
111, 7	" " " " ends ornamented,	67
112, 8½	" " " " with brass angel heads,	89
113, 9¾	" " " " " "	1 35
114, 11¼	" " " " " "	1 90
115, 13	" " " " " "	2 95
116, 15	" " " " " "	4 25
117, 19	" " " " " "	6 40
118, 21	" " " " " "	8 00

WOODEN CRUCIFIXES (ROUND) WITH "PLASTIQUE" FIGURES.

Nos. 211-216.
Nos. 231-235.
Nos. 220-224.

Nos.		Per doz.
211, 8¼	inches high, with round stand and bone tips,	$3 45
212, 9½	" " " " " "	4 70
213, 10½	" " " " " "	6 50
214, 12	" " " " " "	10 25
215, 14	" " " " " "	12 25
216, 15	" " " " " "	15 50

WOODEN CRUCIFIXES (ROUND) WITH BONE FIGURES.

Nos.		Per doz.
231, 8¼	inches high, with round stand and bone tips,	$7 60
232, 9½	" " " " " "	8 00
233, 10½	" " " " " "	10 00
234, 12	" " " " " "	12 00
235, 14	" " " " " "	16 00

WOODEN CRUCIFIXES (ROUND) WITH BRONZE FIGURES.

Nos.		Per doz.
220, 8¼	inches high, with round stand and bone tips,	$
221, 9½	" " " " " "	9 10
222, 10½	" " " " " "	11 25
223, 12	" " " " " "	15 00
224, 14	" " " " " "	20 00

PRICES IN CURRENCY.

WOODEN CRUCIFIXES WITH WOODEN FIGURES.

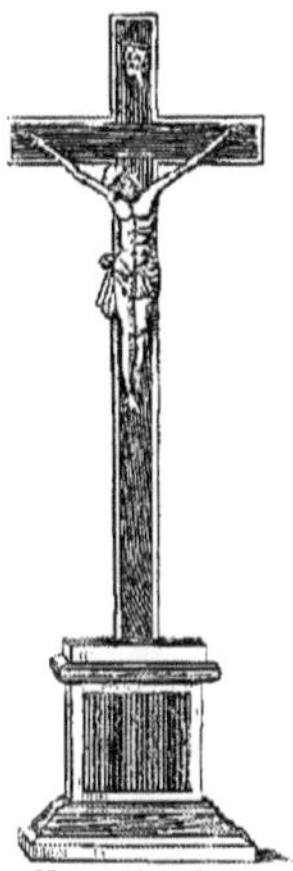

Nos. 190 and 193.

Nos.	Size of Cross, inches.	Size of Figure, inches.		Per piece.
182,	12,	4,	Plain black cross and stand,	$0 47
183,	15,	5,	" " "	70
187,	18,	6,	" " "	78
195,	13½,	4,	Plain black cross and stand,	70
196,	22,	7,	with Mater Dolorosa, .	1 40
197,	16,	5,	Plain black cross and stand, with Mater Dolorosa, and St. John, . . .	97
198,	19,	6,	Plain black cross and stand, with Mater Dolorosa, St. John, and Mary Magdalene,	2 00
170,	13,	4,	Polished cross and stand,	1 45
171,	15½,	5,	" " "	1 85
172,	18,	6,	" " "	2 20
190,	15,	4,	Inlaid cross and stand, .	1 85
193,	18½,	6,	" " "	2 70
199,	15,	4,	Gothic carved cross and stand,	2 30
200,	18½,	5,	" " "	3 90

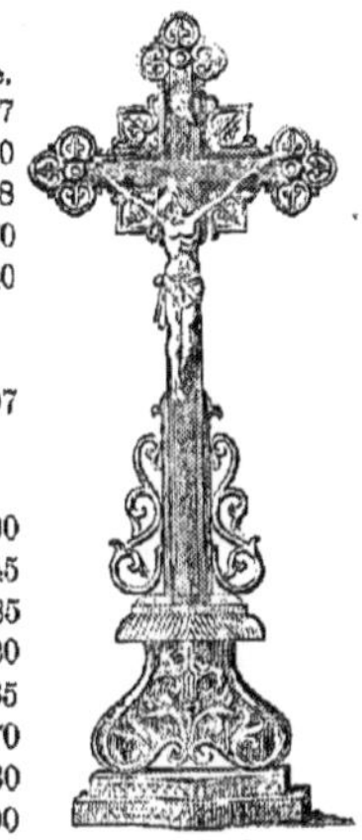

Nos. 199 and 200.

WOODEN CORPUS CHRISTI,

In different sizes and qualities, ... from 50 cents to $6 00

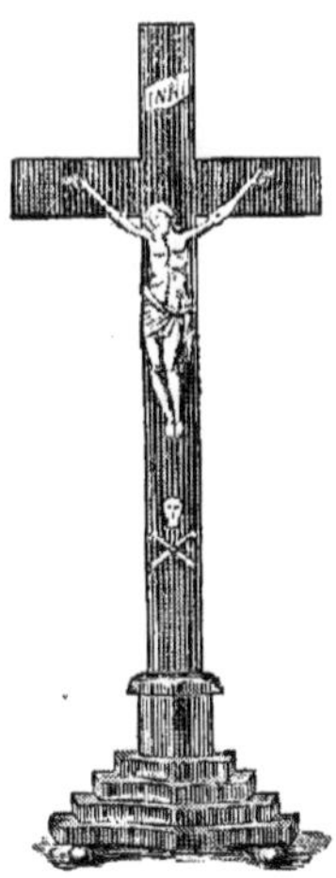

WOODEN CRUCIFIXES WITH IVORY FIGURES.

Nos.	Size of Cross, inches.	Size of Figure, inches.		Per piece.
201,	16,	4¼,	With fine black polished stand, . . .	$ 6 40
202,	18,	5½,	" " " " . .	10 25
203,	20,	6½,	" " " " . . .	15 50
206,	27,	8,	" " " " . .	26 00
208,	16,	4,	Extra fine carving, stand and cross, . .	13 25
209,	18,	5,	" " " " . .	19 50

PRICES IN CURRENCY.

WOODEN CRUCIFIXES WITH CHINA FIGURES.

Nos.			Per doz.
241,	7 inches high, with china stand, gilt,		$1 85
242,	12 " " "		3 55
243,	16 " " "		5 50
244,	18 " " "		8 00
245,	20 " " "		9 20
246,	16 " "	Holy Water Font and statue, gilt,	14 00
247,	24 " "	statue of Mater Dolorosa, "	17 00
248,	13½ " "	Holy Water Font and statue, "	23 00

WOODEN CRUCIFIXES WITH BRONZE FIGURES.

Nos.		Per piece.
151,	19 inches high, flat cross, square stand, figure 6 inches long,	$3 05
152,	21 " " " " 7 "	3 45
153,	24 " " " " 9 "	5 30
154,	30 " " " " 12 "	8 00
161,	16¼ " " without stand, " 6 "	2 40
162,	18½ " " " " 7 "	2 65
163,	21½ " " " " 9 "	4 25
164,	26 " " " " 12 "	6 40
155,	24 " black walnut cross and stand, figure 9 "	3 50
156,	24 " " " gilt, " 9 "	4 00
165,	21 " " " without stand, " 9 "	3 00
166,	21 " " " gilt, " " 9 "	3 25

WOODEN CRUCIFIXES WITH SILVER-PLATED FIGURES AND ORNAMENTS.

Nos.		Per doz.
251,	11¼ inches high, round cross and half round stand, black, polished,	$6 50
252,	11¼ " " and stand, "	7 30
253,	14 " " " " with ornaments,	8 70
254,	14 " " " " richer "	11 00
255,	13 " flat cross, square stand, black, polished,	8 20
256,	15½ " " " "	8 90

CARVED CRUCIFIXES WITH FINE IVORY FIGURES.

To hang or with stand, from 19 to 30 inches high, the figures varying from 6 to 10 inches in length, all carved in superior style, at prices per piece, . . . from $25 00 to $90 00

PRICES IN CURRENCY.

BENZIGER BROTHERS, NEW YORK AND CINCINNATI.

HOLY-WATER FONTS.

Nos.		Per doz
1,	Plain china, Crucifixion,	$ 1 5
2,	" Our Saviour on the Globe,	4 5
4,	" Blessed Virgin with Infant Jesus,	2 2
6,	" Virgin Mary,	3 0
31,	" Crucifixion, large,	2 2
32,	" " small,	1 5
42, 43. 44, 45,	Pewter,	1 8
51,	China, gilt and ornamented, Cross,	1 8
52,	" " " Crucifix,	1 4
53,	" " " Queen of Heaven and Angels,	4 0
54,	" " " Blessed Virgin in Chapel,	5 7
55,	" " " Queen of Heaven,	9 7
61,	" " " " "	3 1
62,	" " " " "	3 5
63,	" " " Crucifix,	3 7
64,	" " " " with Angels,	3 7
65,	" " " "	3 9
66,	" " " Blessed Virgin,	5 4
71,	Biscuit, white, Cross,	9
72,	" " Immaculate Conception,	2 3
73,	" " Virgin with Infant Jesus,	2 7
74,	" " Immaculate Conception,	2 7
75,	" " St. Joseph,	2 7
76,	" " Immaculate Conception,	3 6
77,	Glass, Crucifix or " "	3 6
56,	To stand, china, gilt, and ornamented, Chapel,	5 7
57,	" " " Nativity,	29 0
58,	" " " Christ in the Sepulchre,	39 0

HOLY-WATER FONTS OF REAL BRONZE,

ON BLACK MARBLE PLATES.

Nos.				Per piec
101,	Angel representing Faith,	13½	inches high,	$16 5
102,	Our Lord at Mount Olivet,	13½	"	16 5
103,	Christ bearing the Cross,	13½	"	16 5
104,	Madonna, after Murillo,	13½	"	16 5
105,	Crucifixion with surrounding leaves,	13½	"	16 5
106,	"	13½	"	16 5
107,	Crucifixion, (without holy-water font,)	14½	"	33 6

BISCUIT STATUES.

Nos.	Height, inches.		Per piece.
401.	2½,	Immaculate Conception, Virgin with child, or St. Joseph,	$0 01
402,	3¼,	" " " "	06
403,	4⅛,	" " " "	12
404,	4⅞,	" " " "	19
405,	5¼,	" " " "	24
406,	7¼,	" " " "	50
407,	8,	" " " "	57
408,	9⅛,	" " " "	76
409,	10,	" " " "	95
410,	11¼,	" " " "	1 90

Nos.	Height, inches.		Per piece
411,	13¾,	Immaculate Conception, Virgin with child, or St. Joseph,	$3 8
412,	15,	" " " "	4 9
413,	17,	" " " "	
414,	20,	" " " "	
415.	8½.	St. Patrick,	4
416,	7½.	St. Bridget,	4
417,	7,	Our Lady of Lourdes,	4
418,	8,	" " "	8
419,	9⅜,	" " "	1 2
420,	10⅞,	" " "	1 5

PRICES IN CURRENCY.

SMALL COLORED STATUES FROM MUNICH

Finished in Artistic Style.

Nos.		Height.	Per piece.
01,	Queen of Heaven, with Infant Jesus,	10½ inches.	$3 60
02,	" " " " "	7¾ "	2 15
03,	Immaculate Conception, Praying,	7¾ "	2 15
04,	" " "	11 "	3 60
05,	Mary Mother of Grace,	9 "	3 10
06,	Our Mother of Sorrows,	13 "	4 15
07,	St. Joseph with Lily,	10½ "	3 60
08.	Infant Jesus standing on Globe with Palm,	10 "	3 60
09,	Infant Jesus lying down,	7 "	4 55
10.	Our Saviour, Ecce Homo,	10½ "	3 60
11,	" lying in the Sepulchre,	12 "	5 40
12,	" in His Resurrection,	10½ "	4 15
13,	" "	8 "	2 90
14,	The Good Shepherd,	11 "	4 55
15,	Mater Dolorosa seated with the body of our Saviour,	14 "	9 00
16,	Adoring Angel, kneeling,	6 "	3 35
17,	" " standing,	8 "	3 10
18,	St. Margaret,	9 "	3 10
19,	" Ann with Mary,	10 "	4 10
20,	" Anthony of Padua,	10 "	3 60
21,	" Leonard,	10 "	3 60
22,	" Aloysius,	10 "	3 60
23,	" Teresa,	10 "	3 60
24,	" Elizabeth,	10 "	3 60
25,	" Catharine,	10 "	3 60
26,	" Peter,	10 "	3 60
27,	" Paul,	10 "	3 60
28,	" Sebastian,	10 "	3 60
29,	" Vincent of Paul,	10 "	3 60
30,	" George,	10 "	3 60
31,	" Michael,	10 "	3 60
32,	" Francis of Assisium,	10 "	3 60
33,	" " Xavier,	10 "	5 20
34,	" Nicholas,	10 "	3 60
35,	" Clare,	10 "	3 60
36,	" Crescentia,	10 "	3 60
37,	" Barbara,	10 "	3 60
38,	" John Nepomucene,	10 "	3 60
39,	" Patrick,	11 "	4 15
60,	" Florian,	10 "	3 60
61,	The Holy Family,	13½ "	9 00

STATUES OF IMITATION BRONZE.

Nos.		Height.	Per piece.
30,	Immaculate Conception,	6½ inches.	$1 80
31,	"	7 "	2 00
32,	"	9½ "	2 50
33,	"	9¾ "	2 65
34,	"	12 "	4 00
35,	"	15 "	6 00
36,	St. Joseph with Infant Jesus,	7½ "	2 20
37,	" " "	9¾ "	3 40

PRICES IN CURRENCY.

BENZIGER BROTHERS, NEW-YORK AND CINCINNATI.

CHRISTMAS FIGURES OF THE HOLY FAMILY.

Made of Papier-Mache and Colored.

Nos.			Height.	Per s
11,	9 figures :	Infant Jesus in the Crib, Blessed Virgin, St. Joseph, the 3 Holy Magi, Shepherd, Ox and Ass,	3 inches.	$1 (
12,	8 "	Infant Jesus in the Crib, Blessed Virgin, St. Joseph, the 3 Holy Magi, Ox, and Ass,	4½ "	1 7
13,	17 "	Infant Jesus in the Crib, Blessed Virgin, St. Joseph, the 3 Holy Magi, 2 Arabians, 5 Shepherds, Ox, Ass, and 2 Sheep,	4½ "	3 (
14,	8 "	Infant Jesus in the Crib, Blessed Virgin, St. Joseph, the 3 Holy Magi, Ox, and Ass,	6 "	2 :
15,	19 "	Infant Jesus in the Crib, Blessed Virgin, St. Joseph, the 3 Holy Magi, 1 Arabian, 5 Shepherds, Ox and Ass, 2 Sheep, 1 Horse, and 2 Camels,	6 "	4 (
16,	8 "	Infant Jesus in the Crib, Blessed Virgin, St. Joseph, the 3 Holy Magi, Ox, and Ass,	6½ "	3 1
17,	18 "	Infant Jesus in the Crib, Blessed Virgin, St. Joseph, the 3 Holy Magi, 1 Arabian, 1 Knight, 4 Shepherds, Ox and Ass, 2 Horses, 1 Camel, and 1 Sheep,	6½ "	6 (
18,	8 "	Infant Jesus in the Crib, Blessed Virgin, St. Joseph, the 3 Holy Magi, Ox, and Ass,	12 "	11 2
19,	17 "	Infant Jesus in the Crib, Blessed Virgin, St. Joseph, the 3 Holy Magi, 4 Shepherds, 1 Angel, Ox and Ass, 3 Sheep, and 1 Shepherd's Dog,	12 "	16 0
20,	21 "	Infant Jesus in the Crib, Blessed Virgin, St. Joseph, the 3 Holy Magi, 2 Angels, 1 Arabian, 6 Shepherds, Ox and Ass, 3 Sheep, and 1 Shepherd's Dog,	12 "	19 0
21,	13 "	Blessed Virgin with the Infant Jesus on her lap, St. Joseph, the 3 Holy Magi, Ox and Ass, 1 Shepherd, 3 Sheep, and 1 Shepherd's Dog.	17 "	41 0

Nos.		Per doz.	Nos.	Per do.
26,	Shepherd,	$2 05	36, Sheep,	$0 5
27,	"	2 55	37, "	6
28,	"	3 40	38, "	8
29,	" with sheep,	2 40	39, "	1 2
30,	" "	3 40	40, "	2 4
31,	" "	5 10		

25, Stable—27 inches high, 23 inches wide—containing: Holy Infant in the Crib, Mary, St. Joseph, Ox, and Ass, ... per piece, $27 0

TOY ALTAR FURNITURE.

Pewter ware, in sets, containing: Ostensorium, Chalice, Ciborium, Candlesticks, Crucifix Censer with Boat, Sanctuary Lamp, Cruets, etc., etc.

No. 1, small, $1.05; No. 2, medium, $2.05; No. 3, large, $3.10.

PAPER WEIGHTS.

Nos.		Per doz
1,	French Glass, round, with religious pictures, small,	$8 2
2,	" " " " large,	12 2
3,	" cut, " " small,	12 2

A variety of larger and finer paper weights on hand.

PRICES IN CURRENCY.

No. 17. No. 15 No. 14.

PORCELAIN MEDALLIONS, WITH GILT BRASS FRAMES.

The Paintings are burnt in Porcelain.

os.		Height.	Per piece.
1,	To stand.	3 inches,	$1 25
2,	"	3 "	1 25
3,	"	3¼ "	1 30
4,	"	3½ "	1 35
5,	" or hang,	1⅝ "	1 35
6,	" "	4¾ "	1 90
7,	" "	4¼ "	1 90
8,	" "	5 "	4 35
9,	Enameled, with doors; to stand or hang,	4¼ "	5 20
0,	To stand or hang,	7¾ "	6 00
1,	Holy Water Fout, enameled, to hang,	10½ "	7 00
2,	With doors, to stand,	7¾ "	8 10
3.	Enameled, with doors, to stand or hang,	6 "	8 70
4.	With doors, " "	9 "	9 80
5,	Holy Water Font, enameled, " "	11 "	11 00
6,	On silk velvet, enameled, to hang,	12 "	11 50
7,	Enameled, with doors, to stand or hang.	9 "	17 50

A fine variety of other patterns always on hand; very large choice in frames, as well as in ie paintings.

PLASTIC MEDALLIONS.

eligious Subjects, with oval glasses, in brass frames, per doz.,	from $1 00	to $3 00
" " " in wooden " " piece,	" 1 50	to 25 00

PRICES IN CURRENCY.

SMALL SANCTUARY LAMPS.

Nos.		Per piece
31,	12 inches long, brass, varnished, plain glass, to hang,	$0 7
32,	14 " " " " " " "	8
33,	14 " " " " colored glass, "	1 1
34,	22 " " " " " " "	3 1
35,	36 " " " " " " "	11 2
36,	7 " high, " " plain glass, to stand,	7
37,	7 " " " " colored glass, "	1 0
38,	7½ " " " " plain " "	8
39,	7½ " " " " colored " " with cover,	1 3
40,	9 " " " " plain " "	1 4
41,	9 " " " " colored " " with cover,	2 0

SMALL BRASS CANDELABRAS.

Nos.		Per pair
12,	14½ inches high, varnished, for 3 lights,	$13 5
13,	14½ " " " " 4 "	15 5
14,	10½ " " " " 3 " with onyx ornaments,	7 0
15,	15 " " " " 3 " " "	16 0
16,	16½ " " " " 4 " " "	18 0
17,	18 " " " " 4 " " "	22 0

The following Candelabras can also be used as Candlesticks by removing the branches.

Nos.		Per pair
21,	10½ inches high, varnished, for 3 lights, with onyx ornaments,	$5 9
22,	11¼ " " " " 3 " " "	6 2
23,	12½ " " " " 3 " " "	6 5
24,	14 " " " " 3 " " "	7 5
25,	10½ " " " " 4 " " "	7 0
26,	11¼ " " " " 4 " " "	7 3
27,	12½ " " " " 4 " " "	7 6
28,	14 " " " " 4 " " "	8 6
29,	14½ " " " " 4 " " "	9 2
30,	15½ " " " " 4 " " "	9 6
31,	13¾ " " " " 4 " extra fine, with onyx ornaments,	14 5
32,	15 " " " " 4 " " " "	15 0
33,	16 " " " " 4 " " " "	16 0
34,	13¾ " " " " 5 " " " "	17 0
35,	15 " " " " 5 " " " "	17 5
36,	16 " " " " 5 " " " "	18 5
37,	17 " " " " 4 " with lilies, " "	18 50
38,	17 " " " " 5 " " " "	21 00

PRICES IN CURRENCY.

No. 103.

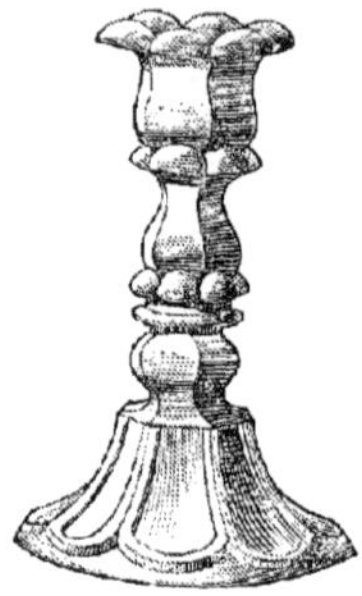

No. 105.

No. 104.

No. 103.–BRITANNIA CANDLESTICKS.

5¼ inches high,	per dozen,	$3 65
7 " "	"	6 00
7¼ " "	"	7 60
0¾ " "	"	10 75
2 " "	"	12 75

No. 104.–GLASS CANDLESTICKS, CRUCIFIX PATTERN.

In Plain or White Glass.

0 inches high,	per dozen,	$4 90
1½ " "	"	6 30

No. 105.–GLASS CANDLESTICKS, PLAIN PATTERN.

In Plain or White Glass.

7 inches high,	per dozen,	$3 50

No. 106.–GLASS CANDLESTICKS, BEADED PATTERN.

In Plain or White Glass.

9 inches high,	per dozen,	$5 60

PRICES IN CURRENCY.

No. 111.—GLASS CANDLESTICKS SILVERED.

4¼ inches high, with silver-plated sockets,	per dozen,	$1 9
6⅛ " " " "	"	2 2
7⅛ " " " "	"	3 4
8¼ " " " "	"	3 8
9¼ " " " "	"	4 8
10 " " " "	"	5 8
11¼ " " " "	"	6 8
8½ " " flowered, with silver-plated sockets,	"	6 0
9½ " " " " " "	"	7 2
10½ " " " " " "	"	8 4

SMALL BRASS CANDLESTICKS.

Nos.		Per pair
1,	4⅝ inches high, varnished,	$0 9
2,	6½ " " "	1 1
3,	7 " " "	1 3
4,	8¼ " " "	1 7
5,	9½ " " "	2 7
6,	10½ " " "	3 2
7,	11½ " " "	3 6
8,	8 " " " with onyx ornaments,	4 2
9,	9½ " " " " " "	4 8
10,	10¼ " " " " " "	5 6
11,	11½ " " " " " " and lilies,	8 2

WAX CANDLES.

Plain White, 1, 2, 3, 4, 6, or 8 to a lb.,	per lb.,	$0 6
Richly Ornamented, 2, 4, or 6 to a lb.,	"	8

WAX TAPERS.

Plain White, 1, 2, 4, or 6 to a lb.,	per lb.,	$0 7
Richly Ornamented, 1, 2, 4, or 6 to a lb.,	"	9

CHRISTMAS CANDLES.

Paraffine, 24, 36, 48, 72, 96, 120, 144 to a lb., . . . per lb., $0 5

ADAMANTINE CANDLES.

Ordinary Quality, 23 cents per lb., or $8.28 per box of 36 lbs.

SUPERIOR QUALITY, MADE EXPRESSLY FOR CHURCH USE, IN WOODEN BOXES OF 36 LBS., (each box containing 6 cartoon boxes of 6 lbs. each,) 4 or 6 to a lb., per lb., 27 cents, . . . $9 7

These latter are the best of the kind in the market, and have been adopted by a grea many churches.

N.B.—The weight for candles, as adopted by manufacturers generally, is 14 ounces t the pound.

PRICES IN CURRENCY.

WREATHS AND BOUQUETS.

FOR FIRST COMMUNICANTS.

No. 10.

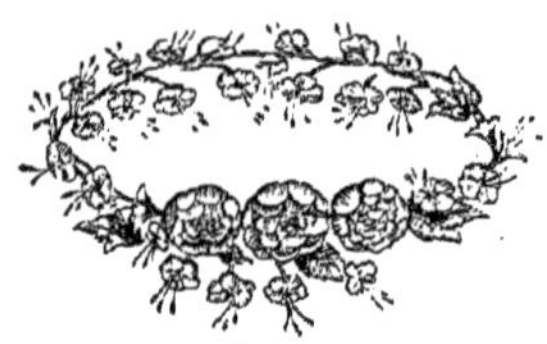

No 15.

WREATHS, (Muslin.)

		Per Dozen.
o. 8.	White blossoms and green leaves	$1 20
" 9.	One white rose, white blossoms and green leaves	1 50
" 10.	Three " roses, " " " " "	1 80
" 11.	Three white camellias, white blossoms and green leaves	3 00
" 12.	Three white roses, double row of white blossoms and green leaves	3 60
" 13.	Three white roses and double row of small white flowers and green leaves	4 80
" 14.	Myrtle Wreath, (paper, mixed with small white muslin flowers)	4 80
" 15.	Three white roses and a triple row of small white May-flowers and green leaves	7 20

Wreaths of richer styles, up to $15 per dozen, made to order.

BOUQUETS, (Muslin.)

		Per Dozen.
o. 1.	One rose, white or red, with green leaves	$ 60
" 2.	One camellia, white or red, three white blossoms and green leaves	75
" 3.	One rose, white or red, five white blossoms and green leaves	1 00
" 4.	One camellia, white or red, five white blossoms and green leaves	1 20
" 5.	One rose, white or red, sixteen small flowers and green leaves	1 50
" 6.	" " " " with many fine blossoms and green leaves	2 10
" 7.	One large camellia, white or red, and many white blossoms, with green leaves, (mostly used with the candles)	2 10

N.B.—White bouquets are mostly used by girls, and red bouquets by boys.

PRICES IN CURRENCY.

BENZIGER BROTHERS, NEW-YORK, AND CINCINNATI, O.

RELIGIOUS PICTURES.

A COLLECTION OF PRINTS TO BE FRAMED, 8x12.

TWO PRINTS ON ONE SHEET.

Size of print, 7x10; size of paper, 11x16½.

PRICE, WITH BLACK BORDER, PLAIN, PER 100 SHEETS, ... $2.00
" " " COLORED, " " 4.45

500 and 501, ...The Sacred Heart of Jesus. The Sacred Heart of Mary.
502 " 503, ...Ecce Homo. Mater Dolorosa.
504 " 505, ...St. Joseph. The Blessed Virgin Mary.
506 " 507, ...Jesus the Children's Friend. Jesus, Mary, and St. John the Baptist.
508 " 509, ...St. John the Baptist. The Redeemer.
510 " 511, ...Two Angels with Ensigns of the Passion.
512 " 513, ...Infant Jesus. The Blessed Virgin with Lily.
514 " 515, ...Infant Jesus with Cross. St. John the Baptist.
516 " 517, ...The Good Shepherd. Immaculate Conception.
518 " 519, ...Ecce Homo. Mater Dolorosa.
520 " 521, ...Christ and His Disciples. Mary and Virgins.
522 " 523, ...Infant Jesus with Chalice. Infant Jesus with Globe.
524 " 525, ...Virgin and Child. St. Joseph with Child.
526 " 527, ...The Sacred Heart of Jesus. The Sacred Heart of Mary.
528 " 529, ...Our Lady of the Rosary. Our Lady of the Scapular.
530 " 531, ...St. Peter. St. Paul.
532 " 533, ...St. Anthony of Padua. St. Aloysius.
534 " 535, ...St. Catharine. St. Ann and the Blessed Virgin.
536 " 537, ...St. Barbara. St. Margaret.
538 " 539, ...St. Rose. St. Apolonia.
540 " 541, ...St. Veronica. St. Francis de Chantal.
542 " 543, ...St. Teresa. St. Elizabeth.
544 " 545, ...St. Anthony of Padua. St. James the Greater.
546 " 547, ...St. George. St. Michael.
548 " 549, ...St. Martin. St. Nicholas.
550 " 551, ...St. Francis Xavier. The Beloved Disciple.
552 " 553, ...Infant Jesus with Grapes. Infant Jesus with the Commandments.
554 " 555, ...Jesus Christ with Chalice. Institution of the Blessed Sacrament.
556 " 557, ...Jesus on the lap of his Mother. Jesus bearing the Cross.
558 " 559, ...St. Nepomucen. St. Stanislas Kostka.
560 " 561, ...Jesus, Mary, and St. John the Baptist. Blessed Virgin with Infant and Lamb
562 " 563, ...Sacred Heart of Jesus and Mary. Our Mother of Help.
564 " 565, ...The Holy Family. The Death of St. Joseph.
566 " 567, ...The Queen of Heaven. Guardian Angel.
568 " 569, ...Madonna della Sedia. Blessed Virgin with Infant and Lily.
570 " 571, ...The Blessed Virgin praying. St. Rosalia.
572 " 573, ...St. Ursula. St. Helene.
574 " 575, ...St. Agnes. St. Philomene.
576 " 577, ...St. Cecilia. St. Magdalene.
578 " 579, ...St. Frederick. St. Francis of Sales.
580 " 581, ...St. Patrick. St. Bridget.
582 " 583, ...St. Francis of Assisium. St. Bridget.
584 " 585, ...St. Lawrence. St. Stephen.
586 " 587, ...St. Benedict. St. Bernard.
588 " 589, ...St. Alphonsus Liguori. St. Henry.
590 " 591, ...St. Charles Borromeo. St. John Berchmans.
592 " 593, ...St. Louis. St. Augustine.
594 " 595, ...St. John the Baptist. Infant Jesus with Globe.
596 " 597, ...St. Aloysius Gonzaga. St. William.
598 " 599, ...St. Ignatius Loyola. St. Boniface.
600 " 601, ...Praise ye the Lord. Agony of our Lord.
602 " 603, ...St. Vincent de Paul. St. Florian.
604 " 605, ...St. Luke. St. Matthew, } the four Evangelists.
606 " 607, ...St. John. St. Mark, }
608 " 609, ...St. Joseph. The Blessed Virgin and St. John.
610 " 611, ...St. Genevieve. St. Gertrude.
612 " 613, ...The Holy Family. St. Ann, the Blessed Mary and Joachim.
614 " 615, ...St. Cunigunda. St. Hedwig.
616 " 617, ...The Good Shepherd. The Nativity.
618 " 619, ...St. Stanislaus. St. Casimir.

Prices in Currency.

BENZIGER BROTHERS, NEW-YORK AND CINCINNATI.

Religious Pictures.

A COLLECTION OF PRINTS,

SUITABLE FOR FRAMING.

Size of Print, 8 x 11; *Size of Paper*, 11½ x 16¼; *To be Framed*, 10 x 14.

RICE, WITH COLORED BORDER, PLAIN, PER 100....... ....$2 90
" " " " COLORED, PER 100.... .... 4 45

(A.) SAINTS.

St. John the Baptist.
" Charles Borromeo.
" Ursula.
" Margaret.
" Veronica.
" Agnes.
" Elizabeth.
" Alphonsus Liguori.
" Francis of Assis
" Nicholas.
" Louis.
" Rosalie.
" John the Evangelist.
" Barbara.
" James.
" Helen.
" Magdalene.
" Frederick.
" Ignatius of Loyola.
" Francis of Sales.
" Martin.
" Peter.
" Rose of Lima.

32 St John Nepomucen.
34 " Francis Xavier.
35 " Laurence.
36 " Paul.
37 " Anthony.
38 " George.
39 " Henry.
40 " Patrick.
41 " Stephen.
42 " Teresa.
43 " Michael.
44 " Catharine.
47 " Apollonia.
49 " Anthony of Padua.
51 " Bridget.
52 " Stanislaus Kostka.
61 " Aloysius Conzaga.
63 " Cecilia.
64 " Augustine.
66 " Joseph.
68 Blessed Nicholas of Fluhe.
79 St. John the Baptist.

85 St. Benedict.
89 " John Berchmanns.
90 " John the Baptist.
94 " Bernard.
98 " William.
102 " Andrew.
103a " Gertrude.
103b " Joseph.
104 " Vincent of Paul.
105 " Francisca of Chantal.
106 " Philomene.
107 " Florian.
108 " Joseph with Infant Jesus.
111 " Luke.
112 " Mark.
113 " Matthew.
115 " Agatha.
116 " Gregory.
117 " Genevieve.
118 " Justina.
119 " Wendelin.
120 " Mathias.

(B.) DIFFERENT SUBJECTS.

Mary, Jesus, and St. John.
The Mother of God.
Follow Me.
The Good Shepherd.
Our Mother of Sorrows.
Ecce Homo.
I am the Bread of Life.
Agony of our Lord.
Blessed are they that Mourn.
Mother of Our Saviour.
The Sacred Hearts of Jesus and Mary.
Mary Conceived without Sin.
Our Hope is in the Cross.
Mother of our Saviour.
Follow Me.
Our Hope is in the Cross.
Holy Virgin of Virgins.
Death of St. Joseph.
The Holy Family.
The Sacred Heart of Mary.
" " " " Jesus.
Blessed are the Pure of Heart.
The Mother of Sorrows.
Mother Most Amiable.
Keep My Commandments.
Such is the Love of God for the World
The Queen of Heaven.
Jesus the Children's Friend.

73 Mother of Our Saviour.
74 St. Anne and the Blessed Virgin.
75 The Guardian Angel.
76 The Queen of Martyrs.
77 Behold the Man.
78 In the Cross is Eternal Life.
80 Jesus, Infinite Goodness.
81 Come to Me All.
82 Jesus, our Way and our Life.
83 The Food of Angels.
84 Meek and Humble of Heart.
86 Abide in My Love.
87 Jesus, Mary, and St. John.
88 My Burden is Light.
91 Praise ye the Lord.
92 I am the Bread of Life.
93 Our Salvation is in Thy Hands.
95 My Soul doth Magnify the Lord unto me.
96 I am the Mother of Pure Love.
97 Faith, Hope, Charity.
99 Suffer Little Children to come unto Me.
101 Blessed are the Clean of Heart.
109 Sacred Heart of Jesus.
110 " " " Mary.
114 Blessed Virgin Mary.
121 Virgin Mary with Infant Jesus.
122 The Good Shepherd.
123 Our Lady of the Scapular.

PRICES IN CURRENCY.

BENZIGER BROTHERS, NEW-YORK AND CINCINNATI.

Religious Pictures.

A Large Collection of Prints, to be Framed, 14 × 10.

Size of paper, 16½ × 11½; size of drawing, 11 × 8.

Price per 100, plain, $3.00; colored, $1.60.

a. Saints (busts only).

5 St. Adelaide.
15 " Susannah.
20 " Elizabeth.
24 " Peter.
25 " Paul.
28 " Matthew,
29 " Mark,
30 " Luke,
31 " John, } the four Evangelists.
32 " Theresa.
33 " Margaret.
34 " Anthony of Padua.
40 " Catharine.
41 " Aloysius of Gonzaga.
48 " Amelia.
56 " Magdalene.
57 " Canigunda.
81 " Walpurga.
84 " Crescentia.
91 " Barbara.
92 " Regina.
99 " Rochus.
132 " Agatha.
138 " Helena.
146 " Francisca.
147 " John Nepomucen.
155 " Ursula.
157 " Christina.
171 " Agnes.
173 " Rupert.
199 St. Stanislaus of Kostka.
200 " Dorothea.
201 " Frederick.
205 " Gerard.
207 " Theodore.
210 " Vincent.
220 " Ann.
221 " Joseph.
232 " Sophia.
242 " Philomena.
313 " Genevieve.
328 " Victoria.
329 " Apolonia.
330 " Othilia.
331 " Cecilia.
332 " Johanna (Jane).
333 " Monica.
334 " Rosalia.
335 " Justina.
336 " Josepha.
337 " Stephen.
338 " Laurent.
340 " Henry.
342 " William.
436 " Augustin.
439 " Gregory.
443 " Clare.
520 " Sybilla.
521 " Christopher.
534 " Herman.

b. Saints (whole figures).

10 St. Veronica.
13 " Francis Xavier.
37 " George.
38 " Michael.
83 " Bernard.
98 " Anthony (the hermit).
108 " Sebastian.
110 " Charles Borromeo.
123 " James (Jacobus).
124 " Philip.
135 " Martin.
158 " Andrew.
162 " Leonard.
168 " Ludgerus.
208 " Valentine.
223 " Ignatius Loyola.
227 " Wenceslaus.
239 " Florian.
249 St. Francis Seraph.
250 " Anthony of Padua.
251 " John.
271 " Matthias.
272 " Nicholas.
339 " Benedict.
341 " Wendelin.
359 " Valentine.
361 " Gertrude.
362 " Victor.
391 " Joseph.
392 " Mary.
400 " Stanislaus.
457 " Blasius (Blase).
458 " Dominick.
512 " Francis of Assisium.
513 " Hubert.
698 " Boniface.

Prices in Currency.

Religious Pictures.

c. Different Prints (whole figures.)

7 The good Samaritan.
21 Jesus on the Cross.
82 The Image of Christ, with medallion border.
85 Immaculate Conception.
114 The Holy Trinity.
115 The Coronation of Mary.
125 Jesus on the Cross with St. John, the blessed Virgin and Mary Magdalene.
139 The World has been redeemed by His Blood.
202 St. Joseph, the Blessed Virgin and the infant Jesus.
206 Holy Mary, conceived without sin.
209 The seven words of Jesus.
228 Death of the Just. (tranv.)
229 Death of the Sinner. (transv.)
230 Christ walks on the Sea.
233 The Sixtinian Madonna.
238 The Guardian Angel.
252 Our Mother of Sorrows.
253 Our Lady of Help. (with Crown)
263 Veneration of the Blessed Virgin.
268 St. John the Baptist.
269 The Good Shepherd.
273 The Image of our Lord.
275 Ave Maria.
276 Our Lady of Help.
277 " " " Good Council.
278 " " " Protection.
281 The Passion of Our Lord.
282 The 5 Wounds of Our Lord.
293 The Way of the Cross.
316 Memorial.
427 The 14 Holy Helpers.
477 The 3 Holy Magi (kings.)
485 Our Lady of Help (tree.)
492 The Holy Eucharist.
493 Agnus Dei.
522 The Holy Corpse of Jesus on the lap of His Mother

d. Different Prints (busts only.)

22 Our Saviour.
23 St. John the Baptist.
54 Sacred Heart of Jesus } large
55 " " " Mary } large
56 Ecce Homo.
59 Mater dolorosa.
60 Jesus Christ.
61 Mater amabilis.
70 Our Redeemer.
71 The Blessed Virgin in prayer.
72 The Good Shepherd.
73 Jesus bearing the Cross.
169 The Redeemer of the World.
170 St. John the Baptist.
211 Our Lady of Help.
212 St. Joseph.
367 Our Lady of the Scapular.
368 " " " " Rosary.
516 Sacred Heart of Jesus } small.
517 " " " Mary } small.
2507 Christ blessing Bread and Wine.
2508 Mater amabilis.
2513 Infant Jesus.
2514 St. John the Baptist.

e Different Prints with Lithographed Borders.

This Series of Prints contains over a hundred different representations, mostly similar to those given above under letter *a*.

f. Different Prints with Border representing the Way of the Cross, or the 15 Mysteries.

11 Our Lord of Angels.
51 Sacred Heart of Jesus.
52 " " " Mary.
78 The Mysteries of the Blessed Virgin.
79 The Holy Eucharist.
285 Jesus dies on the Cross.

g. Prints representing Scenes from the Holy Scriptures.

8 The Baptism of Jesus Christ.
9 Flight to Egypt.
18 The Nativity of our Lord (transv.)
19 The Sepulchre of our Lord, at Jerusalem (transv.)
26 The Assumption.
39 Jesus at Mount Olivet.
46 Jesus taken down from the Cross.
86 The last Supper, (transv.)
98 Father, in thy hands I commend my spirit!
113 Jesus sentenced to Death by Pilate, (transv.)
143 The Marriage Feast at Cana. (transv.)
165 Birth of Christ, (transv.)
184 The Resurrection.
185 The Ascension.
191 The Annunciation
231 Jesus laid in the Holy Sepulchre.
298 The Sermon on the Mountain, (transv.)
299 Jesus weeps over Jerusalem, (transv.)
300 Jesus entering Jerusalem, (transv.)
301 Jesus blessing the Children, (transv.)
523 Jesus the Children's friend.
524 St. Mary (la belle Jardiniere).
613 The Rock of the Church.

h. 2 Prints on One Sheet.

116 The Holy Eucharist and Agnus Dei.
119 Christ on the Cross and our Lord of Angels.
184 St. Joseph and St. Mary.
266 Sacred Heart of Jesus and Mary.
440 The Japanese Martyrs.
471 The Saviour of the World and St. John the Baptist.
3 Sacred Heart of Jesus and Mary } with lithogr. border.
4 The Redeemer & St. John the Bapt. } with lithogr. border.
27 St. Joseph and Our Lady of Help } with lithographed border.
47 Our Saviour and Mater Amabilis
127 Ecce homo and Mater dolorosa
163 St. Aloysius and St. Anthony
460 " Margaret and St. Elizabeth
462 " Anne and St. Therese
467 " Martin and St. George
469 " Barbara and St. Catharine
470 " Joseph and the Blessed Virgin

and a number of other representations.

Prices in Currency.

Religious Pictures.

i. 4 Prints on One Sheet.

A collection of about fifty different representations.

k. Prints in Sets

74, 75, 76, 77 — 4 Scenes of the Life of Joseph of Egypt (transv.)

187, 188, 189, 190 — 4 Scenes of the Life of St. Genevieve (transv.)

383, 384, 385, 386 — 4 Scenes from the Parable of the Prodigal Son. (transv.)

Collection of Lithographs to be Framed 18x13.

Size of paper, 20x14 ; size of drawing, 11½x9.
Price : p. piece, colored, 13 Cts.

701 St. Joseph.
702 Our Lady of Help.
703 St. Patrick.
704 Jesus bearing the Cross.
706 Jesus at Mount Olivet.
707 It is consummated.
711 The Immaculate Conception.
712 Descent from the Cross.
713 Jesus laid in the Holy Sepulchre.
714 Adoration of the 3 Holy Magi (kings).
715 The Marriage of the Holy Virgin.
717 The Offering of Jesus in the Temple.
718 Golgatha.
719 Our Lord of Angels.
721 Ecce homo.
722 Mater dolorosa.
726 The Holy Trinity.
727 Coronation of Mary.
728 Christ blessing bread and wine.
729 Mater amabilis.
735 Jesus, Mary and Joseph.
736 Holy Trinity (new).
737 Coronation of B. Virgin (new).
738 Sixtinian Madonna.
739 The last Judgment.
742 Christ walking on the sea.
743 Flight to Egypt.
749 The 7 words of Jesus.
758 The 7 sacraments.
790 St. Bridget.
818 St. Michael.
823 Father, in thy hands I commend my spirit.
830 Assumption.
847 The Exaltation of the Cross.
848 St. Joseph.
849 Our Lady of Help.
863 The Holy Place of the Birth of Christ.
864 The Holy Sepulchre at Jerusalem.
866 The good Shepherd.
868 Help, O Lord, I am sinking.
885 Ecce Homo. } large.
886 Mater Dolorosa }
887 St. James (Jacobus).
888 St. Philip.
894 Veneration of the Holy Virgin.
906 Christ on the Cross, with St. John, the Bl. Virgin and Mary Magdalene.
908 Sacred Heart of Jesus.
910 " " " Mary.
935 St. Gertrude.
1007 St. Vincent of Paul.
1008 St. Cecilia.
1028 Jesus Christ.
1029 Mater amabilis.
1060 The marriage at Cana (transv.)
1061–1074 The 14 Stations of the Cross (14 Lithographs
1084 St. Anthony of Padua.
1085 St. Therese.
1087 The Queen of Angels.
1108 Jesus at the Pillar.
1113 At the Holy Sepulchre.
1114 The Sermon on the Mountain.
1115 Jesus weeps over Jerusalem (transv.).
1116 Christ enters into Jerusalem (transv.).
1117 Jesus blessing the Children (transv.).
1118 Sacred Heart of Jesus.
1119 " " " Mary.
1120 The Lord's Supper (transv.).
1121 Jesus dies on the Cross.
1122 Faith, Hope and Charity.
1123 Christ bearing the Cross.
1124 Our Lady of Sorrows.
1125 Jesus Christ }
1126 St. Mary } busts in oval borders.
1127 Ecce Homo }
1128 Mater dolorosa }
1129 St. Catharine.
1130 " Elizabeth.
1131 " Margaret.
1132 " Barbara.
1133 " Ann.
1134 " Magdalene.
1135 " Aloysius of Gonzaga.
1136 " Francis Xavier.
1149 The Redeemer of the World } in oval borders.
1150 St. John the Baptist }
1151 " George.
1152 " Martin.
1155 The World has been redeemed by His Blood.
1170 The Resurrection.
1171 The Ascension.
1174 The Holy Night.
1178 The last moments of Jesus.
1181 The Annunciation.
1182 The Baptism of Jesus Christ.
1183 St. Peter.
1184 " Paul.
1228 " Matthew }
1229 " Mark } the 4 Evangelists, whole figures.
1230 " Luke }
1231 " John }

Prices in Currency.

Religious Pictures.

Collection of Lithographs to be Framed, 24 x 18.

Size of paper, 27 x 20; size of drawing, 17 x 13.
Price per piece, plain, 30 cts.; colored, 60 cts.

1 *St. Joseph.*
2 St. Theresa.
3 Jesus Dying on the Cross.
6 St. Anthony of Padua.
7 Our Lady of Sorrows.
8 Our Lady of Mount Carmel.
10 The Immaculate Conception.
11 Our Lady of the Rosary.
13 The Good Shepherd.
16 The Queen of Angels.
17 The Baptism of our Lord.
18 St. Ann.
19 St. Louis of Gonzaga (St. Aloysius).
20 The Nativity.
21 Sacred Heart of Jesus.
22 Sacred Heart of Mary.
23 St. Mary Magdalene.
24 Descent from the Cross.
25 The Crucifixion (Redeemer).
27 The Assumption.
28 The Holy Family.
29 St. Michael the Archangel.
30 The Guardian Angel.
32 Death of the Just Man.
34 Death of the Sinner.
38 St. John the Baptist.
39 Ecce Homo.
40 Mater Dolorosa.
41 St. Nicholas.
42 St. Patrick.
48 St. Catharine.
49 The Holy Trinity.
50 Jesus, Mary, and Joseph.
51 Mary, Mother of God.
52 St. Francis of Paul.
53 St. August st
54 St. Francis of Assisium.
55 St. Peter.
56 Mater Amabilis.
59 The Ascension.
60 St. Anthony of Padoua.
62 The Saviour of the World.
66 St. Francis Xavier.
67 Calvary (The Crucifixion).
69 The Resurrection.
71 St. Vincent de Paul (full length).
72 St. Aloysius (bust).
73 St. Rose of Lima.
74 St. Philomena.
75 St. George.
76 St. Catharine of Sienna.
77 St. Paul.
78 St. Barbara.
79 The Annunciation.
81 St. Ignatius of Loyola.
84 St. Joachim.
85 St. Bridget.
86 St. Lawrence.
88 St. Thomas of Aquin.
90 St. Dominick.
92 St. Charles Borromeo.
99 St. James.
101 Infancy of Jesus.
102 St. Isidore.
104 St. Sebastian.
105 The Good Shepherd.
106 The Good Shepherdess.
109 The Transfiguration.
110 Raphael's Madonna.
118 St. Stanislaus of Kostka.
121 St. Alphonsus Liguori.
122 St. Helena.
123 St. Clara.
125 St. Elizabeth.
127 St. Cecilia.
128 St. Margaret.
132 St. John Nepomucen.
134 St. Francis de Sales.
140 St. Stephen.
141 St. Veronica.
150 St. Martin.
152 Sacred Heart of Jesus (full length).
153 Sacred Heart of Mary (full length).
155 St. Frederick.
158 Our Lady of Lourdes.
159 Our Lady of Perpetual Succor.

Collection of Colored Lithographs on Black Ground with Gilt Borders, to be Framed, 16 x 12.

Size of paper, 17½ x 12; size of drawing with gilt border, 11 x 8½.
Price per sheet, 15 cts.

The Saviour of the World.
St. John the Baptist.
Jesus blessing Bread and Wine.
Mater Amabilis.
Ecce Homo, } Busts.
Mater Dolorosa, } Busts.
St. Joseph, } Busts.
Our Lady of Help, } Busts.
Sacred Heart of Jesus, } Busts.
Sacred Heart of Mary, } Busts.
Christ on the Cross.
The Immaculate Conception.
The Good Shepherd.
Jesus bearing the Cross.
Jesus at Mount Olivet.
Jesus dies on the Cross.
Jesus laid in the Holy Sepulchre.
The Resurrection.
The Ascension.
Our Lord of Angels.
The Blessed Virgin praying to the Infant Jesus.
The Holy Trinity.
The Coronation of the Blessed Virgin.
It is consummated.
Jesus Christ, } Busts.
The Blessed Virgin, } Busts.
The Infant Jesus.
Crucifixion.
Descent from the Cross.
Jesus, the Children's Friend.
Sermon on the Mountain.
Jesus weeps over Jerusalem
Jesus entering Jerusalem.
Queen of Angels.

Prices in Currency.

St. Matthew, St. Mark, St. Luke, St. John — the four Evangelists.

Ecce Homo, Mater Dolorosa, S. Heart of Jesus, " " " Mary, Jesus Christ, Blessed Virgin, — Heads.

The holy Place of the Nativity.
" " Sepulchre at Jerusalem.
Mary, St. Joseph and Infant Jesus.
Guardian Angel.
Sixtinian Madonna.
Come unto me all, &c.
Angelic Salutation.
The 7 last Words of Jesus.
I am the Vine.

Collection of Colored Lithographs on Black Ground, with Gilt Borders, to be Framed 19x14.

Size of paper, 20x14½ ; size of drawing, 12x9
Price : p. piece, 23 cts.

St. Joseph.
Our Lady of Help.
It is consummated.
The Immaculate Conception.
Our Lord of Angels.
At the Holy Sepulchre.
Ecce Homo.
Mater dolorosa.
St. James (Jacobus).
St. Philip.
The Holy Trinity.
The Coronation of the Blessed Virgin.
Christ blessing bread and wine.
Mater amabilis.
The Sacred Heart of Jesus.
" " " " Mary.
Jesus dies on the Cross.
Jesus blessing the Children (transv.)
St. Martin
St. Anthony of Padua.
St. Therese.
The Saviour of the World.
St. John the Baptist.
The Lord's Supper (transv.)
" Holy Night.
" Resurrection.
" Ascension.
" Assumption.
" Annunciation.
" Baptism of Christ.
St. Paul.
St. Peter.
Jesus bearing the Cross.
Jesus taken down from the Cross.
Jesus laid in the Holy Sepulchre.
Christ on the Cross, with St. John and his Bl. Mother.
The Adoration by the 3 Holy Magi.
" " of the Blessed Virgin.
" Transfiguration.
St. Bridget.
" Cecily.
" Vincent of Paul.
" Michael.
St. Gertrude.

The Holy Trinity.
" Coronation.
St. Joseph.
Our Lady of Help.
Jesus entering Jerusalem (transv.)
The Sermon on the Mount (transv.)
Jesus weeps over Jerusalem (transv.)
Christ bearing the Cross.
The sorrowful Mother.

Jesus Christ, St. Mary, Ecce homo, Mater dolorosa — busts, in oval borders.

St. Catharine.
" Elizabeth.
" Margaret.
" Barbara.
" Ann.
" Magdalene.
" Aloysius of Gonzaga.
" Francis Xavier.
" George.
The Holy Place of the Birth of Christ.
" Holy Sepulchre at Jerusalem
" Sixtinian Madonna.
" Queen of Angels.
" Sacred Heart of Jesus.
" " " " Mary.
Jesus at Mount Olivet.
The Marriage Feast at Cana (transv.)
" World has been Redeemed by His Blood.
" seven Words of Jesus.

St. Matthew, " Mark, " Luke, " John, — the 4 Evangelists, whole figures.

" Patrick.
The last Moments of Jesus Christ.
" Exaltation of the Cross.
" last Judgment.
" Flight to Egypt.
" Good Shepherd.
" Way of the Cross (14 lithographs.)

Collection of Colored Lithographs on Black Ground, with Gilt Borders to be Framed 28x20.

Size of paper, 28½x20 ; size of drawing, 21x13.
Price : per piece, 70 cts.

The Sacred Heart of Jesus.
" " " " Mary.
Christ laid in the Holy Sepulchre (transv.)
Christ on the Cross.
The Coronation of Mary.
" Holy Trinity.
Christ dying on the Cross.

Ecce homo.
Mater dolorosa.
The Holy Family.
St. Mary.
Jesus Christ.
The Lord's Supper (transv.)
Descent from the Cross.

Prices in Currency.

Collection of Imitation Photographs, with Embossed Gilt Borders.

IN TWO SIZES. { To be framed 10x8. Price: per 100, $4.35.
To be framed 12x10. Price: per 100, $7.60.

esus blessing bread and wine
Mater amabilis.
The Holy Trinity.
The Coronation of the blessed Virgin.
Come to me All, &c.
Behold the handmaid of the Lord.
The World has been redeemed by His Blood.
Jesus dies on the Cross.
It is consummated.
The Resurrection.
The Ascension.
The seven Words of Jesus.
St. Joseph and Mary with the Infant Jesus.
The Holy Eucharist.
Agnus Dei.
The Holy Corpse of Jesus on the Lap of His Mother.
Jesus appears to Mary.
The image of Jesus.
Jesus at Mount Olivet.
The Guardian Angel.
The Annunciation.
The Baptism of Jesus Christ.
I am the Vine.
Sixtinian Madonna.
The Immaculate Conception.
The Lord's Passion.
The 5 Wounds of our Lord.
The Exaltation of the Cross.
Jesus taken down from the Cross.
St. Mary (la belle Jardinière).
Jesus the Children's friend.
Jesus entering Jerusalem.
Jesus weeps over Jerusalem.
The Sermon on the Mountain.
The 3 holy Magi (kings)
Transfiguration.
Mary, Queen of Angels.
The Saviour of the World.
St. John the Baptist.
Jesus bearing the Cross.
Our sorrowful Mother.
St. Joseph.
Our Lady of Help.
Ecce homo.
Mater dolorosa.
The Sacred Heart of Jesus.
" " " " Mary.
Mater amabilis.
Ecce Homo, } large.
Mater dolorosa, } large.
The sacred Heart of Jesus, } large.
" " " " Mary, } large.
Christ } large.
Mater amabilis } large.

St. John of Nepomuc.
" Barbara.
Our Lady of the Rosary.
" " " Scapular.
St. Magdalene.
" Stanislas.
" Mathias.
" Agatha.
" Rosalia.
" Francisca.
" Elizabeth.
" Catharine.
" Ann.
" Barbara.
" Therese.
" Aloysius.
" Anthony.
" George.
" Martin.
" Francis Xavier.
" Margaret.
" Christina.
" Helena.
" Ursula.
" James (Jacobus).
" Philip.
" Peter.
" Paul.
" Michael.
" Andrew.
" Henry.
" Valentine.
" Wenzeslaus.
" Sebastian.
" Vincent of Paul
" Gebhard.
" Florian.
" Gertrude.
" Clare.
" William.
" Josephine.
" Johanna (Jane).
" Crescentia.
" Walpurga.
" Nicholas.
" Sebastian.
" Cecily.
" Dorothy.
" Cunegund.
" Amalia.
" Rochus.
" Christopher.
" Josepha.
" Agnes.
" Victoria.
" Augustine.
" Monica.

St. Apollonia.
" Philomena.
" Veronica.
" Regina.
Christmas.
Jesus dies on the Cross.
Jesus laid in the Holy Sepulchre.
Our Saviour.
Blessed Virgin praying.
Mary with Infant Jesus.
Annunciation (Murillo.)
The tributary Pence.
It is consummated.
Our Lord of Angels
The Japanese Martyrs.
The Holy Place of the Birth of Christ.
The Holy Sepulchre at Jerusalem.
The Holy Trinity.
The Coronation of the Bl. Virgin.
The Vision of St. Francis of Assisium.
" " " " Felix.
" " " " Anthony of Padua.
Notre Dame of France.
St. Francis of Regis.
Flight to Egypt.
Infant Jesus.
The good Samaritan.
Help, o Lord, I am sinking.
The disbelieving Thomas.
St. Joseph.
Our Lady of Help (With Crown).
Madonna della Sedia.
St. Catharine with Infant Jesus.
Thou art Peter, and upon this Rock, &c.
Jesus insulted.
Jesus scourged.
Procession at the opening of the Council in Rome, Dec. 8th 1869.
Jesus and the Woman of Samaria.
St. Charles Borromeo.
" Ignatius of Loyola.
" Lawrence.
" Stephen.
" Matthew } Evangelists.
" Mark } Evangelists.
" Luke } Evangelists.
" John } Evangelists.

Prices in Currency.

New Collection of 40 Lithographs, Medallion Form, with Narrow Gilt Border, to be Framed 24x18.

Size of paper, 24½x19 ; size of drawing, 15x12.
Price : per piece, plain, 63 cts.; colored, $1.00.
colored, with black ground, $2.00.

1 Jesus Christ our Lord.
2 The Holy Virgin.
3 The Sacred Heart of Jesus.
4 " " " " Mary.
5 St. Paul.
6 " Ann.
7 " Magdalene.
8 " Francis of Assisium.
9 " Peter.
10 " Joseph.
11 " John the Baptist.
12 " James (Jacobus).
13 " Blasius (Blase).
14 " Dominic.
15 " Aloysius of Gonzaga.
16 " Rosa of Lima.
17 " Ignatius of Loyola.
18 " Raymund.
19 Our Lady of Mount Carmel.
20 " " " Sorrows.
21 Our Lady of Mercy.
22 St. Thomas.
23 " Philomena.
24 " Catharine.
25 The Holy Family.
26 St. Helena.
27 " Stanislas of Kostka.
28 " Johanna (Jane).
29 " Bridget.
30 " Patrick.
31 Ecce Homo.
32 The Immaculate Conception.
33 Christ carrying the Cross.
34 The Virgin of the Chair.
35 St. Vincent of Paul.
36 The Guardian Angel.
37 The Angel Raphael and Tobias.
38 The Assumption.
39 Jesus taken down from the Cross.
40 St. Anthony of Padua.

Collection of Large Studies at 2 Crayons; to be Framed 26x20

Size of paper, 26x20½ ; size of drawing, 26x20½.
Price : per piece, tinted, 75 cts.; colored, $1.25 cts.
colored, to imitate oil-painting, 3.00.
" " " " with black ground, $3.00

3 St. Martin.
4 The Immaculate Conception.
6 " Queen of Heaven.
9 " Assumption.
10 " Ascension.
17 Christ dying on the Cross.
20 The Virgin of the Chair.
21 Our Mother of Sorrows.
23 The Mother of Our Saviour.
24 St. Francis Xavier.
25 The Immaculate Conception.
27 The Good Shepherd.
29 St. Peter.
30 " Paul.
31 " Francis of Sales.
33 The Immaculate Conception.
34 Ecce Homo,
36 The Virgin of the Grape.
37 St. Joseph.
39 Christ at the Pillar.
40 St. Mary (La belle Jardinière.)
41 The little St. John.
42 " Sacred Heart of Jesus.
43 " " " " Mary.
44 Christ carrying the Cross.
46 St. Michael.
48 " Anthony.
51 " Magdalene.
52 Father, forgive them.
54 The Education of the Blessed Virgin.
56 St. Anthony of Padua.
58 The Nativity of Our Lord.
60 St. Genevieve.
61 " Charles Borromeo.
62 " Vincent of Paul.
64 " Louis (king of France.)
65 The Caresses of the Infant Jesus.
66 Virgin, Immaculate Mother.
71 Jesus carrying the Cross.
75 Mater amabilis.
77 The Virgin of Roses.
80 " Sacred Heart of Jesus.
81 " " " " Mary.
82 St. Cecilia.
83 " Therese.
84 Pope Pius IX.
85 Immaculate Conception, with Angels.
86 Jesus taken down from the Cross.
95 The Baptism of Christ.
96 " Flight to Egypt.
97 " Annunciation.
98 " Marriage of the Virgin Mary.
99 " Adoration of the 3 Holy Magi (kings.)
100 St. Augustin and St. Monica.
101 " Stanislas of Kostka.
102 " Aloysius of Gonzaga.
103 The Guardian Angel.
104 " Angel Raphael and Tobias.
105 " Saviour of the World.
106 St. Patrick
107 " Ignatius.
108 The Holy Family.
109 Our Lady of Mount Carmel.
110 Agnus Dei.

Prices in Currency.

Collection of Chromo-Lithographs, with Chromo-Lithographic Border to be Framed 19x14.

Size of paper, 19½x15; size of drawing, 15x10½.
Price, per piece, 15 cts.

St. Gertrude.
" Helena.
" Bernard.
" Margaret.
" Laurent.
" Andrew.
" Augustine.
Sebastian.
Barbara.
" Catharine.
" John the Baptist.
" Paul.
" Francisca.
" James (Jacobus.)
Clare.
" Aloysius.
" William.
" Anthony, the hermit.
" Henry.

St. Magdalene.
" Theodorus.
" Ann.
The Blessed Virgin.
St. Christina.
" Joseph.
" Sophia.
" John of Nepomus.
" Charles Borromee.
" Jane (Johanna.)
" Thereso.
" Cecily.
" George.
" Elizabeth.
" Matthew,
" Mark.
" Luke,
" John,
} the 4 Evangelists.

Stations of the Cross in Lithographs.

N.B.—The sizes given below are those of the drawings without margin.

HEIGHT. Inches.		WIDTH. Inches.			
6	by	9	The Stations of the Cross, plain	per Set	$0.90
6	"	9	" " " " colored	"	1.80
6	"	9	" " " " tinted	"	1.80
6	"	9	" " " " dark ground	"	6.75
8	"	9½	" " " " plain	"	1.80
8	"	9½	" " " " colored	"	3.60
10½	"	8	" " " " plain	"	1 80
.0½	"	8	" " " " colored	"	3.60
8	"	11½	" " " " plain	"	2 25
8	"	11½	" " " " colored	"	4.50
.0	"	14	" " " " plain	"	2.70
.0	"	14	" " " " colored	"	5 40
/6	"	12	" " " " plain	"	6.00
16	"	12	" " " " colored	"	15.00
12½	"	18	" " " " plain	"	6.75
.2½	"	18	" " " " colored	"	13.50
16½	"	25	" " " " plain	"	11.25
16½	"	25	" " " " colored	"	22.50
21	"	27½	" " " " plain	"	13.50
21	"	27½	" " " " colored	"	27.00
24	"	38½	" " " " plain	"	15.75
24	"	38½	" " " " colored	"	31.50

N.B.—The sizes of the following are those of the paper (drawing and margin.)

20	"	14	The Stations of the Cross, colored	per Set	$2.00
20	"	14	" " " " dark ground	"	4.00

Gold, Rosewood or Rustic Frames for Stations and Oil-paintings made to order at the lowest prices.

Stations of the Cross in Oil-paintings and in Oil-prints on Canvass always on hand in different sizes.

Prices in Currency.

CHROMOS IN OIL COLORS.

(Our own publications.)

No.	Size of Chromos. Height, inches.	Width, inches.	On paper.	On pasteboard.	With gilt frame.	With very rich gilt frame
1, Infant Jesus reposing on the Cross,	8¼	11	$0 60	$0 75	$2 00	$2 75
2, The Blessed Virgin, with Jesus and St. John,	8¼	11	60	75	2 00	2 75
3, Three Flying Angels: Faith, Hope, and Charity,	8¼	11	60	75	2 00	2 75
4, The Nativity,	8¼	11	60	75	2 00	2 75
5, Infant Jesus and St. John the Baptist,	11	8¼	60	75	2 00	2 75
6, Regina Angelorum, (Queen of Angels,)	17½	10½	1 00	1 20	2 50	3 75
7, The Crucifixion,	17½	11	1 00	1 20	2 50	3 75
8, Our Lady of the Rosary, with the 15 Mysteries,	9	6	40	55	1 40	2 25
9, Guardian Angel watching over two lost children,	8½	5½	25	40	1 25	2 00

The above Chromos if ordered on paper or on pasteboard will be sent free by mail on receipt of price; if desired in gilt frames, they will be sent by express, and 25 cents extra will be charged for box.

CHROMOS ON PASTEBOARD OR ON STRETCHERS.

		Size	Price
Easter Morning, cross with flowers,		7 × 10¾	$ 1 5
" " " finer,		7 × 10¾	2 2
Anchor and Flowers,		11 × 14	1 0
Crosses in Mats,		11 × 14	7
Cross, Lyre, and Flowers,		14 × 17	3 0
Wreaths for Inscriptions,		14 × 17	3 0
Madonna, after Murillo, fine,		14 × 18	7 5
" " "		20½ × 25½	15 0
Ecce Homo, and Mater Dolorosa,	each,	9 × 12	5
Christ bearing the Cross,		14 × 18	8
Pope Pius IX.,		13 × 17	9
St. Patrick,		16 × 18	9
"		26 × 36	7 0
Madonna Sixtina,		17 × 22½	2 4
The Good Shepherd,		17 × 22½	2 4
Our Lady of Help,		17 × 22½	2 4
Ecce Homo, and Mater Dolorosa,	each,	17 × 22½	2 4
St. Joseph with Child,		17 × 22½	2 4
Blessed Virgin with Child,		17 × 22½	2 4
Holy Family, after Hess,		18 × 24½	2 8
" " Raphael,		22½ × 28	2 8
Sacred Heart of Jesus, and Sacred Heart of Mary,	each,	20 × 26	2 8
St. Aloysius,		18 × 23	2 4
The Ascension,		18 × 23	2 4
The Assumption,		18 × 23	2 4
Madonna della Sedia,		22½ × 28	2 8
Ecce Homo, and Mater Dolorosa,	each,	20 × 26	5 2
Madonna Sixtina,		25 × 35	8 0
Madonna della Sedia,		25 × 30	6 0
Immaculate Conception, after Murillo,		25 × 35	8 0
Pope Pius IX., extra fine,		18 × 24	3 0

CHEAP OIL PAINTINGS.

We keep a large variety of the most desirable religious subjects, mostly Saints, copies after good masters, and executed in fine coloring; of the following sizes:

24½ inches high by 18½ inches wide, at . . . $3 00
30½ " " 24½ " from . . . 5 50 to $11 50

PRICES IN CURRENCY.

BENZIGER BROTHERS, NEW-YORK AND CINCINNATI.

Religious Pictures.

SMALL LITHOGRAPHS, IN PACKS.

Prices per Pack—100 Pictures.

No.	Height. Inches.	Width. Inches.	
2,	4,	2½,	Saints, 68 subjects : Plain, 14 cts. ; in one color, 17 cts. ; colored, 50 cts.
4,	4½,	3⅛,	Saints, 60 subjects : With chromo and gilt borders and gold ground, plain, 77cts. ; colored, $1.50.
6,	3⅝,	2½,	Saints, 38 subjects : Border in one color, plain, 19 cts. ; colored, 77 cts. Border in two colors, plain, 26 cts. , colored, 85 cts. Gilt border, plain, 26 cts. ; colored, 85 cts. Chromo border, plain, 34 cts. ; colored, with gold ground, 94 cts With scalloped edges, 10 cts. extra.
7,	4,	2½,	Saints, 32 subjects : Chromo border, plain, 60 cts. ; colored, $1.25.
9,	4¼,	2¾,	32 Saints, and different religious subjects : Chromo border, plain, 60 cts. ; colored, $1.55. With scalloped edges, 10 cts. extra.
1,	3⅞,	2⅜,	Saints, 41 subjects : Colored border, plain, 24 cts. ; colored, 83 cts. Gilt border, plain, 34 cts. ; colored, 92 cts. Silver border, plain, 34 cts. ; colored, 92 cts. With scalloped edges, 10 cts. extra.
8,	4¼,	2¾,	Saints, 62 subjects : Colored border, plain, 29 cts.; colored, $1.
20,	3¾,	2⅜,	Saints, 42 subjects : Plain, 12 cts.; in one color, 16 cts.; colored, 46 cts.
22,	4¼,	2¾,	Saints, 64 subjects : Tinted, 26 cts. ; colored, 96 cts. Colored border, tinted, 34 cts. ; colored, $1.10. Gilt border, tinted, 46 cts. ; colored, $1.20. With English prayers on back, 10 cts. extra.
27,	3¾,	2⅜,	Saints, 42 subjects : Colored border, plain, 34 cts.; colored, 80 cts With scalloped edges, 10 cts. extra.
35,	4¼,	3,	Saints, 60 subjects : Colored border, tinted, 31 cts. ; colored, 87 cts. With English prayers on back, 7 cts. extra.
43,	5,	2½,	Saints, with a short description of their lives, in German. 120 subjects Plain, 51 cts.
55,	3¾,	2½,	Saints, 32 subjects : Chromo and gilt border, plain, 60 cts. ; colored, $1.20.
56,	3¾,	2½,	Saints, 30 subjects : Chromo and gilt border, plain, 77 cts. ; colored, $1.70.
61,	4¼,	2¾,	32 different religious subjects : Plain, 60 cts. ; colored, $1.55.
63,	5,	3⅜,	Saints, 60 subjects : Colored border, plain, 43 cts. ; colored, $1.15. Gilt border, plain, 60 cts. ; colored, $1.30.
65,	4¼,	2¾,	Saints, with a short description of their lives, in German. 96 subjects Plain, 28 cts.
67,	4¼,	2¾,	64 Saints and different religious subjects : Tinted, 26 cts. ; colored, $1.20.

PRICES IN CURRENCY.

BENZIGER BROTHERS, NEW-YORK AND CINCINNATI.

SMALL LITHOGRAPHS, IN PACKS. (Continued.)

Prices per Pack—100 Pictures.

No.	Height. Inches.	Width. Inches.	
8,	4¼,	2⅞,	Saints, 32 subjects : Chromo and gilt border, plain, 60 cts. ; colored, $1.15.
9,	3,	2⅛,	44 Saints and different religious subjects : Plain, 12 cts. ; in one color, 14 cts. ; colored, 46 cts. Colored border, plain, 17 cts. ; colored, 51 cts. Gilt border, plain, 23 cts. ; colored, 57 cts.
1,	3,	2⅛,	38 Saints and different religious subjects : Colored border, plain, 19 cts. ; colored, 55 cts. Chromo border, plain, 24 cts. ; colored, 63 cts. With scalloped edges, 10 cts. extra.
2,	3,	2,	Saints, etc., 57 subjects : Plain, 11 cts. ; in one color, 12 cts. ; colored, 51 cts. Colored border, plain, 16 cts. ; colored, 57 cts. Gilt border, plain, 23 cts. ; colored, 63 cts. With scalloped edges, 10 cts. extra.
3,	3,	2⅛,	Saints, etc., 56 subjects : Colored border, plain, 16 cts. ; colored, 57 cts. Gilt border, plain, 23 cts. ; colored, 63 cts.

SMALL STEEL ENGRAVINGS, IN PACKS.

Prices per Pack—100 Pictures, assorted.

No.	Size of Paper. Inches.	Size of Engraving. Inches.		
,	9½ × 6¼,	4⅛ × 2⅝,	Saints and different religious subjects,	$3 00
,	7¼ × 4⅝,	4⅛ × 2⅝,	" " " "	2 55
,	9½ × 6¼,	3¼ × 2⅛,	" " " "	2 85
,	6⅜ × 4¼,	3¼ × 2⅛,	" " " "	2 25
,	9½ × 6¼,	2¾ × 1⅞,	" " " "	2 55
,	5 × 3½,	2¾ × 1⅞,	" " " "	2 00
l,	4¼ × 3,	2¾ × 1⅞,	"	1 70

GELATINE PICTURES, IN PACKS.

Prices per Pack—100 Pictures.

No.	Height. Inches.	Width. Inches.		
1,	2⅛,	1¾,	Saints, etc.,	$0 31
2,	3,	2,	" "	41
3,	3¼,	2½,	" "	51
4,	4¼,	3,	" "	70

CHROMOS, IN PACKS.

Prices per Pack—50 Pictures.

No.		
1,	Medium, Saints, with chromo border,	$0 62
2a,	" " " gilt "	45
2b,	" " " red "	34
5,	Large, Saints and Infant Jesus, with gilt border,	77
6,	" Infant Jesus and groups, " " "	94
7,	Small, Saints, with purple border,	31
0,	" Emblems, with flowers, extra fine,	51
1,	Smallest, " " "	34
2,	Small, " extra fine,	51
\,	Medium, Pope Pius IX., with gilt border,	1 20

PRICES IN CURRENCY.

Religious Pictures.

CHROMOS, IN SHEETS.

Saints, different Religious Subjects, Emblems, etc.

2 to 160, Pictures on one sheet,	. . . .	per sheet, from	60 cts.	to 80 cts.
0 to 160, " " "	fine, ...	" "	85 "	to $1 35
0 to 90, " " "	extra fine, ...	" "	$1 50	to 1 80

MOURNING PICTURES.

Prices per Pack—100 Pictures.

lo.	Height. Inches.	Width. Inches.	
8,	4¼,	3,	Lithographs, 20 subjects : Plain, with silver ground, 55 cts.
7,	4½,	3,	Lithographs, 23 subjects : Plain, with silver border, 85 cts.
8,	4½,	3,	Lithographs, 24 subjects : Plain, tinted, 60 cts.
0,	4¼,	2¾,	Lithographs, 25 subjects : Plain, 28 cts.
2,	4¼,	2¾,	Lithographs, 23 subjects : Plain, with black border, 28 cts.
4,	4½,	3,	Lithographs, 20 subjects : Plain, 51 cts. ; silver print, 94 cts.
0,	4⅝,	3,	Lithographs, 18 subjects : Plain, 51 cts.
,	6,	4,	Steel engravings : With black border, $2.30.
,	4⅝,	3,	Steel engravings : With black border, $2.30.

N. B.—The name of deceased, time of death and appropriate text, as may be desired, will be printed on the back for from $2 to $2.50 extra, per pack.

GALLERY OF RELIGIOUS STEEL ENGRAVINGS.

FROM PAINTINGS AND ORIGINAL DRAWINGS OF

I. Paul v. Deschwanden, S. Benz, C. Classen, Th. v. Deschwanden, Maria Ellenrieder, Gebh. latz, Jos. Fuerich, Adam Huber, Lochner, Mignard, P. Molitor, Murillo, Jos. Obwexer, Overbeck, Raphael, Guido Reni, Sassoferato, Schlottauer, Schraudolph, etc.

ENGRAVED BY

Allgeyer, N. Barthelmess, J. M. Baumann, Joh. Burger, Conr. Geyer, Fr. Ludy, Fr. Massau, Chr. Preisel, J. L. Raab, Al. Rohrdorf, Adr. Schleich, N. Stang, J. F. Vogel, etc.

This "Gallery" in large 8vo. (size of paper 9½ × 6⅜ inches) contains the best of our enravings, and is acknowledged by connoisseurs—in France, as well as in Germany—to be the nest and most elaborate collection of small religious engravings ever published.

The "Gallery" contains 38 numbers now, but receives additions from time to time. Price r number (containing 6 engravings), 28 cents.

PRICES IN CURRENCY.

BENZIGER BROTHERS, NEW-YORK AND CINCINNATI.

LACE PICTURES.

SIZES: 48mo., about 3 × 2 inches; 32mo., about 3¾ × 2½ inches.
" 24mo., " 4 × 2¾ " 18mo., " 4½ × 2⅞ "

			48mo.	32mo.	24mo.	18mo.
Lithographs, plain,	per 50 pictures,			32	35	
" " with colored border,	" "			38	47	
" " " gilt "	" "			53	59	
Chromos, "	" "	from	38	50	86	
		to		75		
" with gilt border,	" "	from		1 10	94	1 00
		to				1 20
" " chromo cover, (flowers,)	" "	from	83	1 00	1 25	
		to		1 75	1 50	
" Pope Pius IX.,	" "					1 20
" plain,	per dozen, (13 pictures,)	from	24	21	26	
		to		29	42	
" with gilt border,	" "	from	19	24	34	47
		to	33	47	55	70
" " ground,	" "	from	33	32	41	
		to		36	55	
" " border and chromo cover,	" "		37	36	56	70
" " " (Pope Pius IX.,)	" "				43	51
Steel Engravings, plain,	" "	from	-	32	32	53
		to		47	53	68
" " with gilt border,	" "	from	14	26	33	42
		to		36	47	61
" colored,	" "	from		53	47	57
		to		68	84	1 00
" " " "	" "	from		68	72	90
		to		1 50	1 50	1 50
" " " colored covers,	" "	from			1 15	1 40
		to			1 40	1 60
French Steel Engravings, plain,	" "	from		25	32	45
		to		45	45	66
" " colored,	" "			88		1 15
" " plain, spangled,	" "					1 90
" " colored, "	" "					2 50
" " " extra rich spangled,	"					3 55
" " dressed figures,	"			1 35		2 65
" " for first communicants, (for boys or girls,) plain, colored, spangled, and with dressed figures, per dozen, (13 pictures,)		from		44 cts. to		3 55

STEEL ENGRAVINGS, WITH EMBOSSED BORDER.

Price per Pack—100 pictures.

No.			Price
900,	Saints, (busts,)	4⅜ inches × 2¾ inches,	3 15
901,	Saints and groups,	4¾ " × 2¾ "	3 15
902 and 903,	" "	4½ " × 2⅝ "	2 75
904,	" "	4⅜ " × 2½ "	2 45
905,	Saints,	4½ " × 2½ "	2 35
907,	"	4 " × 2⅝ "	1 85
908,	Infant Jesus,	4 " × 2½ "	2 45
1001,	Saints, with lace border,	3¼ " × 2¼ "	65
1002,	" " "	4½ " × 2¾ "	1 25

Nos. 904 and 908, with English prayers, cost 21 cents more per pack.

PRICES IN CURRENCY.

COMMUNION CARDS.

WITH ENGLISH OR GERMAN TEXT.

No.	Height, inches.	Width, inches.		Per dozen.
6,	10,	7½,	The Holy Communion, plain,	$ 17
6,	10,	7½,	" " colored,	58
0,	11,	8,	" " plain, with colored border,	36
0,	11,	8,	" " chromo, " "	84
1,	11,	8,	" " plain, " "	30
1,	11,	8,	" " chromo, " "	84
3,	11½,	8,	" " tinted,	26
3,	11½,	8,	" " colored,	76
3,	11½,	8,	" " plain, with colored border,	32
3,	11½,	8,	" " chromo, " "	84
4,	11,	8,	" " chromo, with chromo border,	1 30
5,	11,	8,	" " steel engraving, " "	76
5,	11,	8,	" " chromo, " "	1 30
7,	11½,	8½,	" " plain, with colored border,	31
7	11½,	8½,	" " colored, " "	1 10
8,	11½,	8½,	Christ blessing the bread and wine, engraving,	80
8,	11½,	8½,	" " " " " colored, with colored border,	68
9,	11,	8½,	The Holy Communion, plain, with colored border,	36
9,	11,	8½,	" " chromo, " "	84
0,	11,	8¼,	" " " " tinted "	95
1,	11,	8¼,	" " " " " "	95
2,	12½,	9½,	Christ surrounded with angels, giving Holy Communion to boys and girls; extra fine chromo,	2 05

COMMUNION AND CONFIRMATION CARDS.

No.		
12,	Communion and Confirmation, for boys and girls, extra fine chromo, folio,	$2 55
104,	" " " " " fine steel engraving, large folio,	1 60
032,	" " " " " " " "	1 35
033,	" " " " " " " "	1 60

CONFIRMATION CARDS.

,	Lithograph, plain; size, 11½ × 8½,	61
,	" colored; " 11½ × 8½,	1 30

THE 15 MYSTERIES OF THE HOLY ROSARY.

'o.		
,	15 Lithographs on one sheet, with text in English or German; per sheet, plain,	$ 07
,	15 " " " *plain, with colored border,*	*09*
1,	15 Photographs, with brown lace border, per set,	1 50

THE WAY OF THE CROSS.

'o.		
,	14 Lithographs on one sheet, with text in English or German, per sheet, plain,	$ 07
),	14 Photographs, with brown lace border, per set,	1 50

SAMPLE PACKAGES OF OUR PICTURES AT COST PRICES.

SENT FREE BY MAIL.

'o.		
,	One package, containing 88 samples of the most saleable numbers of our lithographs and engravings, in packs,	$1 00
,	One package, containing 10 samples of our most saleable communion and confirmation cards,	1 00
.	One package, containing 100 samples of our most saleable lace pictures, (steel engravings,)	5 00
,	One package, containing samples of the most saleable small chromo pictures, with lace and without lace,	

PRICES IN CURRENCY.

BENZIGER BROTHERS, NEW-YORK AND CINCINNATI.

CATALOGUE
OF
APPROVED EDITIONS
OF
STANDARD CATHOLIC PRAYER-BOOKS,

PUBLISHED BY

BENZIGER BROTHERS,

NEW-YORK: 311 BROADWAY.

CINCINNATI: VINE STREET.

These books have all been examined by the proper authorities and bear the IMPRIMATUR *of the Most Reverend Archbishop.*

In point of excellence, whether in regard to paper, printing, or binding, they will be found to be unequalled, whilst the prices are as low as any.

In ordering it will only be necessary to give the numbers which designate the particular style and book needed.

A liberal discount from these prices will be made to Dealers, Clergymen, Religious Institutions, and Schools.

Any book in this Catalogue will be mailed free, on receipt of the price.

The Catholic Prayer Book. 18mo. 960 pages.

1001 Cloth, sprinkled edges $1 00
1002 Roan embossed, sprinkled edges 1 25
1003 " " gilt edges 1 75
1003½ " " " clasp 2 00
1004 French mor., gilt centre and edges 2 00
1004½ " " " clasp 2 25
1004¾ " " " rim and clasp 2 75
1005 " " full gilt sides, back, and edges 2 25
1005½ " " " " " clasp 2 50
1006 " " in relief, gilt edges 2 00
1006½ " " " " clasp 2 25
1006¾ " " " " rim and clasp 2 75
1007 Turkey mor., gilt centre and edges 3 50
1007½ " " " " clasp 4 00
1008 " " ant. extra, gilt centre and edges 4 25
1008½ " " " " " clasp 4 75
1008¾ " " " " rim and clasp 5 25
1009 " " super, gilt sides, back, and edges 4 25
1009½ " " " " " " clasp 4 75
1010 " " relief, gilt edges 6 00
1010½ " " " " clasp 6 50
1011 Calf, gilt cross on side, red edges 4 50
1011½ " " " " clasp 5 00
1012 " " " gilt edges 4 50
1012½ " " " " clasp 5 00
1013 Silk velvet, rim and clasp, cross in centre 8 50
1014 " " " rich ornaments 10 00

Ivory sides and back, edges red under gold. $20 to 30 00

Also in a variety of other styles.

The Pious Guide. 18mo. 698 pages.

A "LARGE PRINT" book. With 36 full-page illustrations of the Mass.

1101 Cloth $0 90
1102 Roan embossed, sprinkled edges 1 25
1103 French mor., gilt centre and edges 1 50
1103½ " " " clasp 1 75
1103¾ " " " rim and clasp 2 25
1104 " " full gilt sides, back and edges 1 75
1104½ " " " " " " clasp. 2 00
1105 " paneled, gilt sides, back and edges 3 00
1105½ " " " " " clasp 3 25
1106 Turkey mor., gilt centre and edges 3 00
1106½ " " " clasp 3 50
1106¾ " " " rim and clasp 4 00
1107 " ant. extra, gilt centre and edges 4 00
1107½ " " " " clasp 4 50
1107¾ " " " " rim and clasp 5 00
1108 " super, gilt sides, back and edges 4 00
1108½ " " " " " clasp 4 50
1109 Silk velvet, rim and clasp, ornamental centre 8 00
1110 " " " rich ornaments 9 00

True Piety. 18mo. 576 pages.

1201 Roan embossed, sprinkled edges $1 00
1202 " " gilt edges 1 25
1202½ " " " clasp 1 50
1203 French mor., gilt centre and edges 1 50
1203½ " " " " clasp 1 75
1203¾ " " " " rim and clasp 2 25

No.	Binding	Price
04	French mor., full gilt sides, back and edges	$1 75
04½	" " " " clasp	2 00
05	Turkey mor., gilt centre and edges	3 00
05½	" " " " clasp	3 50
05¾	" " " " rim and clasp	4 00
06	" " ant. extra, gilt centre and edges	3 75
06½	" " " " clasp	4 25
06¾	" " " " rim and clasp	4 75
07	" " super, gilt sides, back and edges	3 75
07½	" " " " " clasp	4 25
08	" " relief, gilt edges	4 50
08½	" " " " clasp	5 00
09	Calf, gilt centre and edges	4 00
09½	" " " clasp	4 50
10	" " red edges	4 00
10½	" " " clasp	4 50
11	Silk velvet, rim and clasp, ornamental centre	8 00
12	" " " " rich ornaments	9 00

he Garland of Prayer. 16mo. 392 pages.

Each page is ornamented with a border of exquisite de-
;n engraved expressly for this book.

No.	Binding	Price
01	French mor., gilt centre and edges	$2 00
01½	" " " " clasp	2 25
01¾	" " " " rim and clasp	2 75
02	Turkey mor., ant., gilt centre and edges	3 00
02½	" " " " " clasp	3 50
02¾	" " " " " rim and clasp	4 00
03	" " " extra, gilt centre and edges	4 00
03½	" " " " " clasp	4 50
03¾	" " " " " rim and clasp	5 00
04	" " super, gilt sides, back and edges	4 00
04½	" " " " " clasp	4 50
05	" " relief, gilt edges	4 50
05½	" " " " clasp	[illegible]
06	" " circuit, flexible, gilt edges	4 50
07	Calf, gilt centre and edges	4 25
07½	" " " clasp	4 75
08	" " red edges	4 25
08½	" " " clasp	4 75
09	" flexible, gilt edges	4 25
10	" " red edges	4 25
11	Silk velvet, rim and clasp, cross in centre, edges red under gold	9 00
12	White silk velvet, rim and clasp, cross in centre, edges red under gold	10 00
	Ivory sides and back, carved or applied ornaments	$20 00 to 40 00

he Catholic Guide. 24mo. 489 pages.

Printed with a RED LINE.

No.	Binding	Price
01	French mor., gilt centre and edges	$1 50
01½	" " " " clasp	1 75
01¾	" " " " rim and clasp.	2 25
02	" " full gilt sides, back, and edges	1 75
02½	" " " " " clasp	2 00
03	Persian calf, illuminated centre, gilt edges	2 00
03½	" " " " clasp	2 25
03¾	" " " " rim and clasp.	2 75
04	Turkey mor., ant., extra, beveled, gilt edges	3 50
04½	" " " " " clasp	4 00
04¾	" " " " " rim and clasp.	4 50
05	" " super, gilt sides, back and edges	3 50
05½	" " " " " clasp	4 00

No.	Binding	Price
1406	Turkey mor., relief, gilt edges	$4 00
1406½	" " " " clasp	4 50
1407	Calf, gilt centre and edges	3 75
1407½	" " " clasp	4 25
1408	" " red edges	3 75
1408½	" " " clasp	4 25
1409	Silk velvet, rim and clasp, gilt edge	6 50
1410	" " " orna. centre	7 00
1411	" " " rich ornaments	8 00

The Key of Heaven. 24mo. 444 pages.

No.	Binding	Price
1501	Cloth	$0 50
1502	Roan embossed, sprinkled edges	0 65
1503	French mor., gilt centre and edges	0 90
1503½	" " " " clasp	1 15
1503¾	" " " " rim and clasp	1 65
1504	" " embossed sides, gilt edges	0 90
1504½	" " " " " clasp	1 15
1504¾	" " " " " rim and clasp	1 65
1505	" " full gilt sides, back and edges	1 15
1505½	" " " " clasp	1 40

The Key of Heaven, with Epistles and Gospels. 24mo. 544 pages.

No.	Binding	Price
1601	Cloth	$0 60
1602	Roan embossed, sprinkled edges	0 75
1603	French mor., gilt centre and edges	1 00
1603½	" " " " clasp	1 25
1603¾	" " " " rim and clasp.	1 75
1604	" " full gilt sides, back, and edges	1 25
1604½	" " " " " clasp	1 50
1605	" " embossed sides, gilt edges	1 00
1605½	" " " " " clasp	1 25
1605¾	" " " " " rim and clasp	1 75
1606	" " paneled sides, full gilt, centre in relief, gilt edges	2 00
1606½	" " " " " " clasp	2 25
1607	Persian calf, illuminated centre, gilt edges	1 50
1607½	" " " " clasp	1 75
1607¾	" " " " rim and clasp	2 25
1608	Turkey mor., gilt centre and edges	2 50
1608½	" " " " " clasp	3 00
1608¾	" " " " " rim and clasp	3 50
1609	" " extra ant., gilt centre and edges	3 00
1609½	" " " " clasp	3 50
1609¾	" " " " rim and clasp	4 00
1610	" " cross, polished mor., in relief, g't edg.	3 50
1610½	" " " " " " clasp	4 00
1611	" " per, full gilt sides, back and edges	3 00
1611½	" " " " " " clasp	3 50
1612	Calf, gilt centre, carmine edges	3 25
1612½	" " " " clasp	3 75
1613	" inlaid centre, carmine edges	3 50
1614	English patent velvet, rim and clasp, gilt edges	2 25
1615	Silk velvet, gilt edges, rim and clasp	6 00
1616	" " " " ornamental centre	6 50
1617	" " " " rich ornaments	7 00
1618	Ivory sides and back, gilt edges, clasp	13 00
1619	" " extra, silver clasp	18 00

The Path to Heaven. 32mo. 239 pages.

Printed with a RED LINE.

No.	Binding	Price
1701	French mor., gilt centre and edges	$1 00
1701½	" " " " clasp	1 25
1701¾	" " " " rim and clasp	1 75

1702 French mor., full gilt sides, back, and edges....$1 25
1702½ " " " " " clasp. 1 50
1703 Persian calf, blind tooling, gilt edges........... 1 00
1703½ " " " " clasp........ 1 25
1703¾ " " " " rim and clasp 1 75
1704 French mor., relief, red edges.................. 1 50
1704½ " " " " clasp............. 1 75
1705 Turkey mor., ant., extra, gilt centre and edges.. 2 50
1705½ " " " " " clasp......... 3 00
1705¾ " " " " " rim and clasp 3 50
1706 " " super, full gilt sides, back, and edges 2 50
1706½ " " " " " " clasp 3 00
1707 " " circuit, gilt edges................. 3 00
1708 Calf, gilt centre and edges...................... 2 75
1708½ " " " clasp................. 3 25
1708¾ " " " rim and clasp........ 3 75
1709 Silk velvet, gilt edges, rim and clasp........... 4 00
1710 " " " " orna. centre 4 50
1711 " " rich ornaments...................... 5 00
1712 Ivory sides and back, gilt edges, clasp..........10 00
1713 " " " extra, silver clasp........13 00

The Manual of Piety. 32mo. 487 pages.

1801 Cloth...$0 50
1802 Roan embossed, plain.......................... 0 65
1803 " " gilt edges....................... 0 75
1803½ " " " clasp................ 1 00
1804 French mor., gilt centre and edges............. 1 00
1804½ " " " " clasp...... .. 1 25
1804¾ " " " " rim and clasp 1 75
1805 " " full gilt sides, back and edges...... 1 25
1805½ " " " " " clasp. 1 50
1806 " " relief sides and back, gilt edges.... 1 25
1806½ " " " " " " clasp 1 50
1806¾ " " " " " rim and clasp 2 00
1807 Persian calf, blind tooling, gilt edges.......... 1 00
1807½ " " " " " clasp........ 1 25
1807¾ " " " " " rim and clasp 1 75
1808 Turkey mor., ant., extra, gilt centre and edges.. 2 50
1808½ " " " " " clasp........ 3 00
1808¾ " " " " " rim and clasp 3 50
1809 " " super, full gilt sides, back, and edges 2 50
1809½ " " " " " " " clasp 3 00
1810 " " circuit, gilt edges................. 3 50
1811 Calf, flexible, gilt centre and edges............. 2 75
1812 " " " " red edges.............. 2 75
1813 Silk velvet, rim and clasp, gilt edges............ 4 00
1814 " " " " " orna. centre.. 4 50
1815 " " " " " rich ornaments 5 00

Manual of the Sodality of the Blessed Virgin Mary. 32mo. 613 pages.

1901 Cloth...$0 50
1902 " extra, bevelled, red edges................ 0 75
1903 French mor., gilt centre and edges............. 1 00
1903½ " " " " clasp......... 1 25
1903¾ " " " " rim and clasp. 1 75
1904 " " full gilt sides, back, and edges.... 1 25
1904½ " " " " " " clasp 1 50
1905 Turkey mor., extra ant., gilt centre and edges.. 2 50
1905½ " " " " clasp.......... 3 00
1905¾ " " " " rim and clasp. 3 50
1906 " " super, full gilt sides, back and edges 2 50
1906½ " " " " " " " clasp 3 00
1907 Calf, flexible, red edges......................... 2 75

The Child of Mary. 32mo. 502 pages

2001 Cloth...$0 5
2002 Roan embossed, sprinkled edges................ 0 6
2003 " " gilt edges...................... 0 7
2003½ " " " clasp................ 1 0
2004 French mor., gilt centre and edges............. 1 0
2004½ " " " " clasp,........ 1 2
2004¾ " " " " rim and clasp. 1 7
2005 " " full gilt sides, back, and edges.... 1 2
2005½ " " " " " " clasp. 1 5
2006 Persian calf, chromo centre, gilt sides, back, & ed. 1 2
2006½ " " " " " clasp. 1 5
2007 Silk velvet, rim and clasp, gilt edges............ 4 0
2008 " " " " orna. centre... 4 5
2009 " " " " rich ornaments. 5 0

The Garden of the Soul. 32mo. 38 pages.

2101 Cloth...$0 4
2102 Roan embossed, sprinkled edges.................0 6
2103 French mor., gilt centre and edges............. 0 8
2103½ " " " " clasp......... 1 1
2103¾ " " " " rim and clasp. 1 6
2104 " " full gilt sides, back, and edges.... 1 0
2104½ " " " " " " clasp 1 2
2105 " " relief sides and back, gilt edges.... 1 0
2105½ " " " " clasp......... 1 2
2105¾ " " " " rim and clasp. 1 7
2106 Persian calf, blind tooling, gilt edges........... 0 8
2106½ " " " " clasp......... 1 1
2106¾ " " " " rim and clasp. 1 6
2107 " " illuminated centre, gilt edges...... 1 2
2107½ " " " " clasp.......... 1 5
2107¾ " " " " rim and clasp. 2 0
2108 Turkey mor., ant., extra, gilt centre and edges.. 2 5
2108½ " " " " " " clasp 3 0
2108¾ " " " " " rim and clasp 3 5
2109 " " super, full gilt sides, back, and edg. 2 5
2109½ " " " " " clasp 3 0
2110 " " relief, gilt edges.................... 3 0
2110½ " " " " clasp............. 3 5
2111 " " circuit, gilt edges.................. 3 0
2112 Calf flexible, gilt centre and edges............. 2 7
2113 " " red edges.............. 2 7
2114 Silk velvet, rim and clasp, gilt edges........... 4 0
2115 " " " " ornamen. centre 4 5
2116 " " " " rich ornaments 5 0
2117 Ivory sides and back, gilt edges, clasp..........10 0
2118 " " " " extra, silver or ivory clasp..13 0

The Angel Guide. 32mo. 320 pages.

With 36 full-page illustrations of the Mass, and 14 of the Stations of the Cross.

2201 Cloth...$0 4
2201½ Imitation cloth, full gilt....................... 0 4
2202 Cloth, inked, chromo centre, gilt edges........ 0 6
2203 French mor., gilt centre and edges.............. 0 7
2203½ " " " " clasp......... 1 0
2203¾ " " " " rim and clasp 1 5
2204 " " full gilt sides, back, and edges ... 1 0
2204½ " " " " " " clasp 1 2
2205 Persian calf, gilt edges......................... 0 7
2205½ " " " clasp...................... 1 0
2205¾ " " " rim and clasp............ 1 5

'he Christian's Guide to Heaven. 32mo. 319 pages.

301 Cloth....................$0 35
302 Roan embossed, sprinkled edges.......... 0 50
303 " " gilt edges.......... 0 65
303½ " " " clasp.......... 0 90
303¾ " " " rim and clasp.......... 1 40
304 French mor., full gilt sides, back, and edges.... 1 00
304½ " " " " " " clasp 1 25
305 Persian calf, chromo cen., gilt sides, back, and ed. 1 00
305½ " " " " " clasp 1 25

'lowers of Piety. Small 32mo. 256 pages.

401 Cloth....................$0 30
402 Roan embossed, sprinkled edges.......... 0 40
403 " " gilt edges.......... 0 50
404 French mor., gilt centre and edges.......... 0 65
404½ " " " " clasp.......... 0 90
404¾ " " " " rim and clasp. 1 40
405 " " full gilt sides, back and edges..... 0 80
405½ " " " " " " clasp 1 00

'he Key of Heaven. 48mo. 239 pages

501 Cloth....................$0 25
502 Roan embossed, sprinkled edges.......... 0 35
503 French mor., gilt centre and edges.......... 0 50
503½ " " " " clasp.......... 0 75
503¾ " " " " rim and clasp. 1 25
504 " " full gilt sides, back, and edges.... 0 75
504½ " " " " " " clasp 1 00
505 " " tuck, gilt centre and edges.......... 0 75
506 Persian calf, blind tooling, gilt edges.......... 0 50
506½ " " " " " clasp.......... 0 75
506¾ " " " " " rim and clasp 1 25
507 Turkey mor., ant. extra, gilt centre and edges... 1 50
507½ " " " " " clasp 2 00
507¾ " " " " rim and clasp 2 50
508 " " super, full gilt sides, back, and edg. 1 50
508½ " " " " " " clasp 2 00
509 Calf flexible, gilt centre and edges.......... 1 50
510 Silk velvet, rim and clasp, gilt edges.......... 3 00
511 " " " " orna. centre... 3 50
512 " " " " rich ornaments 4 00

'he Daily Exercise. 48mo. 192 pages.

601 Cloth....................$0 18
602 Roan embossed, gilt edges.......... 0 40

'he Imitation of Christ. 32mo. 481 pages.

701 Cloth....................$0 50
702 Extra cloth, beveled, red edges.......... 0 65
703 French mor., gilt centre and edges.......... 1 00
704 Turkey mor., ant, extra, gilt edges.......... 2 50
705 " " " super, gilt edges.......... 2 50
706 " " " circuit, gilt edges.......... 3 25
707 Calf antique, red edges.......... 2 75
708 " flexible, gilt edges.......... 2 75

'he Imitation of Christ. Red Line Edition. 18mo. 481 pages.

801 Extra cloth, beveled, red edges....................$1 25
802 French mor., gilt centre and edges.......... 1 75
803 Turkey mor., ant., extra, gilt edges.......... 4 00

2804 Turkey mor., super, full gilt sides, back, and edges $4 00
2805 " " circuit, gilt edges.......... 5 00
2806 Calf flexible, gilt edges.......... 4 50
2807 " stiff boards, red edges.......... 4 50

The Catholic's Pocket Manual of Devotion. 48mo. 240 pages.

With full-page illustrations of the Mass.

2901 Cloth,....................$0 25
2901½ Imitation cloth, full gilt.......... 25
2902 Roan embossed, sprinkled edges.......... 35
2903 French mor., gilt centre and edges.......... 50
2903½ " " " " clasp.......... 75
2904 " " tuck, gilt centre and edges.......... 75

The Little Path to Heaven. 48mo. 288 pages.

With full-page illustrations of the Mass and Stations of the Cross.

3001 Cloth....................$0 30
3001½ Imitation cloth, full gilt.......... 30
3002 Roan embossed, sprinkled edges.......... 40
3003 French mor., gilt centre and edges.......... 60
3003½ " " " " clasp.......... 85
3004 " " full gilt sides, back and edges.... 75
3004½ " " " " " clasp 1 00
3005 Turkey mor., ant., extra, gilt centre and edges. 1 50
3005½ " " " " " clasp.......... 2 00
3005¾ " " " " " rim and clasp 2 50
3006 " " super, full gilt sides, back, and edges 1 50
3006½ " " " " " " " clasp 2 00
3007 Calf, flexible, gilt centre and edges.......... 1 50
3008 Silk velvet, rim and clasp, gilt edges.......... 3 00
3009 " " " " " orna. centre.. 3 50
3010 " " " " " rich ornaments 4 00

Short Prayers for Little Catholics. 48mo. 192 pages.

With full-page illustrations of the Mass.

This book is intended for those children who have not yet made their First Communion, and the Prayers are of a character suited to the intelligence of the youngest reader.

3101 Imitation cloth, full gilt....................$0 25
3102 Cloth, gilt centre.......... 25
3103 French mor., gilt centre and edges.......... 50
3103½ " " " " " clasp.......... 75
3104 Per ian calf, chromo centre, gilt sides, back and ed. 75
3104½ " " " " " " clasp 1 00

The Epistles and Gospels for the Sundays and Principal Festivals throughout the Year. 32mo. 169 pages.

3201 Cloth....................$0 30

Manual of the Holy Family Sodality. 32mo. 480 pages.

3301 Cloth....................$0 50
3302 Roan embossed, gilt edges.......... 75
3303 French mor., gilt centre and edges.......... 1 00
3303½ " " " " " rim and clasp 1 75
3304 " " full gilt sides, back and edges..... 1 25
3304½ " " " " " clasp 1 50

www.ingramcontent.com/pod-product-compliance
Ingram Content Group UK Ltd.
Pitfield, Milton Keynes, MK11 3LW, UK
UKHW021326280426
12143UKWH00013B/112